Building Pathology and Rehabilitation

Volume 16

This book series addresses the areas of building pathologies and rehabilitation of the constructed heritage, strategies, diagnostic and design methodologies, the appropriately of existing regulations for rehabilitation, energy efficiency, adaptive rehabilitation, rehabilitation technologies and analysis of case studies. The topics of Building Pathology and Rehabilitation include but are not limited to - hygrothermal behaviour - structural pathologies (e.g. stone, wood, mortar, concrete, etc…) - diagnostic techniques - costs of pathology - responsibilities, guarantees and insurance - analysis of case studies - construction code - rehabilitation technologies - architecture and rehabilitation project - materials and their suitability - building performance simulation and energy efficiency - durability and service life.

More information about this series at http://www.springer.com/series/10019

J. M. P. Q. Delgado
Editor

Durability of Concrete Structures

Springer

Editor
J. M. P. Q. Delgado
CONSTRUCT-LFC
Department of Civil Engineering
Faculty of Engineering
University of Porto
Porto, Portugal

ISSN 2194-9832 ISSN 2194-9840 (electronic)
Building Pathology and Rehabilitation
ISBN 978-3-030-62824-6 ISBN 978-3-030-62825-3 (eBook)
https://doi.org/10.1007/978-3-030-62825-3

This Springer imprint is published by the registered company Springer Nature Switzerland AG
The registered company address is: Gewerbestrasse 11, 6330 Cham, Switzerland

Preface

The improvement of energy efficiency, comfort, hygrothermal behavior, and health in buildings is based on the application of building physics principles. The construction industry efficiency is attained through processes improvement, life cycle assessment, and smart management and maintenance.

One of the main goals of the European Union is the transition to Circular Economy. Within the Circular Economy package, Construction and Demolition is one of the priority sectors. In fact, the Construction sector and its related activities result in the intensive consumption of raw materials and the production of high amounts of Construction and Demolition Waste. According to European, the construction sector accounted for 73% of the materials extracted and a third of the waste generated. Considering that, one of the main pillars of the circular economy is the efficient use of resources and raw materials, which imply the use of circular products and maximum usage of materials, and solutions to this problem are urgently needed.

The main purpose of this book, *Durability of Concrete Structures*, is to provide a collection of recent research works related to structural stability and durability, service life, reinforced concrete structures, recycled materials, and sustainability with endogenic materials.

The book is divided into eight chapters that intend to be a resume of the current state of knowledge for benefit of professional colleagues, scientists, students, practitioners, lecturers, and other interested parties to network. At the same time, these topics will be going to the encounter of a variety of scientific and engineering disciplines, such as civil, mechanical, and materials engineering.

Porto, Portugal J. M. P. Q. Delgado

Contents

Application of NDT for Installation of Rooftop Solar Power Plants

Prafulla Parlewar

Abstract The application for Non-destructive Test (NDT) for rooftop solar power plants identifies the use of NDT for structural stability to install rooftop solar panels. Solar power plant installation on old buildings may result into structural failure due to weight of solar panel system. So, before installation of solar panels, NDT test can identify the strength of structural members. Some of the common methods used for the diagnosis of old buildings are Rebound Hammer Test (RHT) and Ultrasonic Pulse Velocity Test (UPV). In this chapter, the application of NDT test for the solar power plant installation is investigated though a case study. Then, this research illustrated the RHT application for installation of rooftop solar power plant. A case study of 700 kW solar plant on 15 buildings is illustrated by applying the NDT test. This chapter discusses how this application can be used to diagnose the building before installation of solar power plants?

Keywords Solar power plant · NDT test · Rebound Hammer test · Structural stability

1 Introduction

Our traditional electrical energy requires fossil fuels like coal and natural gas. Most of these sources are dilapidating and causing green house effects. So, it is important to install the roof top solar power plants which can produce green and clean energy. Installation of rooftop solar power plants not only give clean energy but also helps combating green house gases. This has many advantages, however, installation on old building is structurally unsafe. Because in many old building the old roofs are not structurally stable to take the weight of the proposed solar power plant.

Also, particularly in Reinforced Cement Concrete (RCC) structure, the old members like slab, beams and columns are unstable due to deteriorated conditions. The research here looks into various questions. How do the installation of the solar

P. Parlewar (✉)
City Development Corporation (P) Ltd, Mumbai, India
e-mail: citycorporationindia@gmail.com

J. M. P. Q. Delgado (ed.), *Durability of Concrete Structures*, Building Pathology and Rehabilitation 16, https://doi.org/10.1007/978-3-030-62825-3_1

power plant affect the loading condition of the structural members of the building? Which method can be applied for the structural stability of buildings for installation of rooftop solar power plant? How does the solar panels generate the dead and wind load on the roof of the building? Some of these questions are investigated in this chapter to understand the application of Non-destructive Test (NDT) for stability analysis of buildings for installation of solar power plants.

The NDT for the building to understand behevior of the concrete members are generally Rebound Hammer test (RHT), Ultra Sonic Pulse Velocity test (UPV) and load test. RHT is most simple to understand the compressive strength of R. C. C. members. RHT results on the structural member like slab, beam and columns can easily estimate the compressive strength. These can be further compared with loading of the solar panels on the roof of the building. This simple application ensure the installation of the solar power plant.

In this chapter, a case study of 700 kWp solar power plant on 15 old building in Mumbai (Bombay) petrochemical industry is demonstrated for showing application of RHT for rooftop solar power plant. All these building were constructed between the span of 30 to 50 years. Most of these building are petrochemical operation buildings. All these buildings are R. C. C. structure. Due to the environmental conditions in the petrochemical industry, these buildings deteriorated with passage of time. Generally, in petrochemical industries, hazardous and toxic air pollutant affects the concrete members and steel reinforcement. In Mumbai (Bombay), heavy rain are also responsible to deteriorate the concrete roof slabs and other members.

The project included solar module mounted solar panels on 15 building. Roof footprint of most of these buildings were completely replaced with the module mounted solar panels to achieve maximum electrical energy. These module mounted solar panels were single slope 10° tilt angle. These were installed with Mild Steel (M.S.) galvanised iron assembly and a concrete pedestal on the roof slab. This system consist of vertical post, support members, module mounted member, base clamps, L brackets and concrete pedestal.

The NDT was carried on the concrete slab, columns and beams through RHT in 15 building. The results of the RHT indicated that the compressive strength of the concrete members were not sufficient to carry the loads of the solar panels.

2 Solar Power Plant—Physical and Load Profile

The solar power plant include system of solar panels installed on the roof top of the building. One units is of size 992 mm × 1956 mm in six module. These module structure is mounted on the system of vertical post, support member, module mounted member, base clamp, and L brackets, The vertical post is 70 × 40 × 40 ×2 thick Mild Steel (MS) galvanise steel members. These are supported on members of size 70 × 40 × 40 × 2 thick MS galvanised steel. Then they are further supported by base clamp and L bracket of size ISA 100 × 100 × 6 and ISA 60 × 60 × 4 respectively (Figs. 1 and 2).

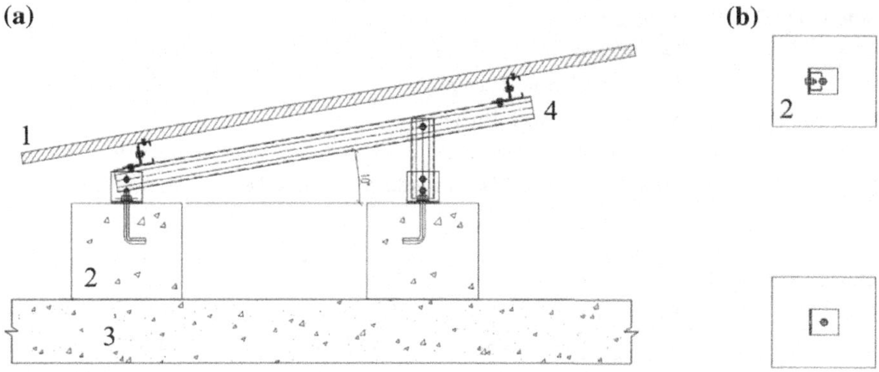

Fig. 1 **a** Section of solar panel system, **b** Plan at concrete pedestal level (1-Photovoltaic Panel, 2-Concrete Pedestal and 3-R.C.C. Slab)

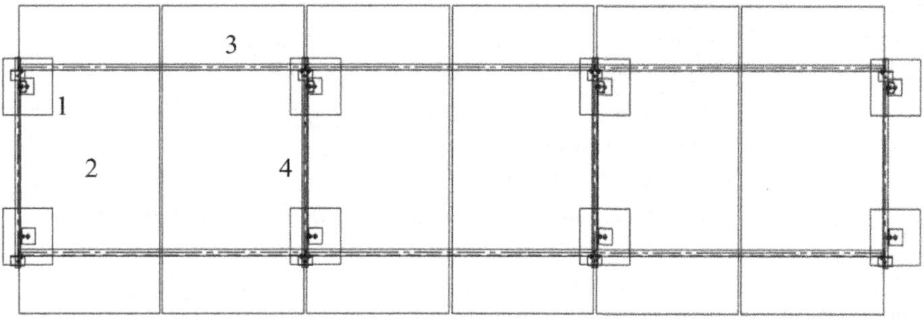

Fig. 2 Plan of module mounted solar power system (1-Concrete Pedestal, 2-Solar Panel, 3-Horizontal support member and 4-Support member between two pedestals)

The uniformly distributed load (UDL) for one solar panel is 0.11 kN/m^2 (Table 1). The wind load on the solar panel affects the load of the building. There are many parameters that influence the values of the drag forces (ground clearance, the inclination and size of the panels etc.); the most important one is the wind direction (Sauca et al. 2019). The designed wind pressure for the solar panel is 0.8402 kN/m^2 (Table 2). The wind pressure will act in upward and downward direction. The UDL in the upward directions will be 0.74 kN/m^2 and the UDL on downward direction will be 0.41 kN/m^2 (Table 3). The concrete pedestal will generate a downward force on the structural member (Table 4). The downward force generated by the pedestal will be 0.92 kN/m^2 (Table 5).

Now, to calculate the force on the solar panel, pressure coefficient is required to be estimates. The pressure coefficient for a mono-slope free roof 10° tilt angle solar panels are: (a) Cp = 0.9000 (upward wind) and (b) Cp = 0.5000 (downward wind).

Table 1 Dead load

S/N	Description	Qty.
1	Size of solar panel (Length)	1.956 m
2	Size of solar panel (Breadth)	0.992 m
3	Total module exposed area	1.940 m^2
4	Dead load (Vertical Downward Direction) (UDL)	0.11 kN/m^2

Table 2 Wind load

S/N	Description	Qty.
1	Base wind speed (Vb)	44.00 m/s
2	Base wind speed (Vb)	158.4 Kmph
3	Risk coefficient factor (K1)	0.9
4	Factors for variation in height (K2)	1.05 kN/m
5	Factors for topography (K3)	1.0
6	Factor for cyclonic regions	1.0
7	Design wind speed (Vz = VbxK1xK2xK3xK4))	1.0
8	Wind directionality factos (Kd)	0.9
9	Area average factor (Ka)	1.0
10	Combination factor (Ka)	0.9
11	Design wind pressure (Pd = KdxKaxKcxPz)	0.8402 kN/m^2

Table 3 Calculation of upward and downward forces

S/N	Description	Qty.
1	Total upward force (Pressure Coefficient × Wind pressure × exposed area)	1.4673 kN
2	UDL on purlin	0.74 kN/m
3	Total downward force (Pressure Coefficient × Wind pressure × exposed area)	0.8152 kN
4	UDL on purlin	0.41 kN/m

Table 4 Pedestal downward force

S/N	Description	Qty.
1	Length	0.35 m
2	Width	0.35 m
3	Height	0.30 m
4	Area	0.03675 m^3
4	Weight (Downward force) of pedestal	0.92 kN

Table 5 Pedestal horizontal force

S/N	Description	Qty.
1	Friction coefficient	0.65 m
2	Effective horizontal force	0.4 kN
3	Frictional resistance of pedestal	1.0 kN

3 Building Profiles

The project includes 15 buildings of R.C.C. constructed during last 30 to 50 years (Figs. 3, 4 and 5). Some of these buildings are plant building and other are the core administration areas. Most of this building are two to three floors. Gener-

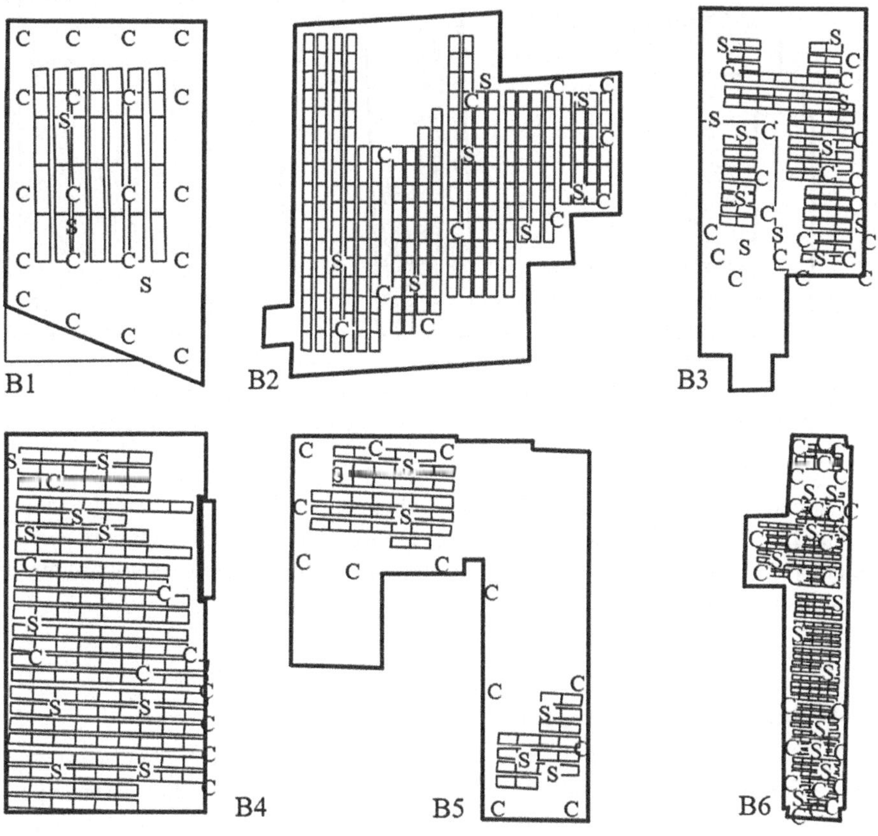

Fig. 3 Roof plan with solar panels showing location of RHT for columns and slabs, (C-Columns and S-Slabs)

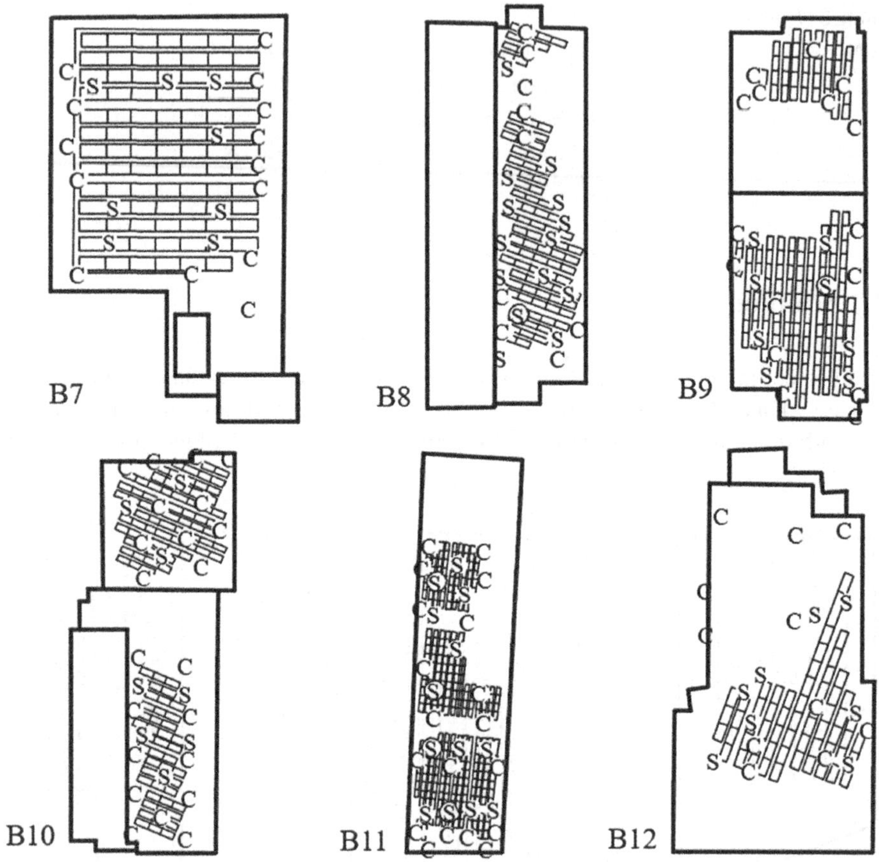

Fig. 4 Roof plan with solar panels showing location of RHT for columns and slabs (C-Columns and S-Slabs)

ally, in petroleum industry the existence of chemicals in the atmosphere deteriorate the building faster than buildings located in the cities. Moreover, the other reason for the deterioration of these buildings are the heavy rain the Mumbai (Bombay) during the monsoon seasons.

4 Rebound Hammer Test

The rebound hammer test as per IS 13311 (Part II 1992), reaffirmed 2004 is principally a surface hardness test (RHT). It works on the principle that the rebound of the elastic mass depends on the hardness of the surface against which the mass impinges (Katalin et al. 2015). RHT gives information on the compressive strength of the concrete.

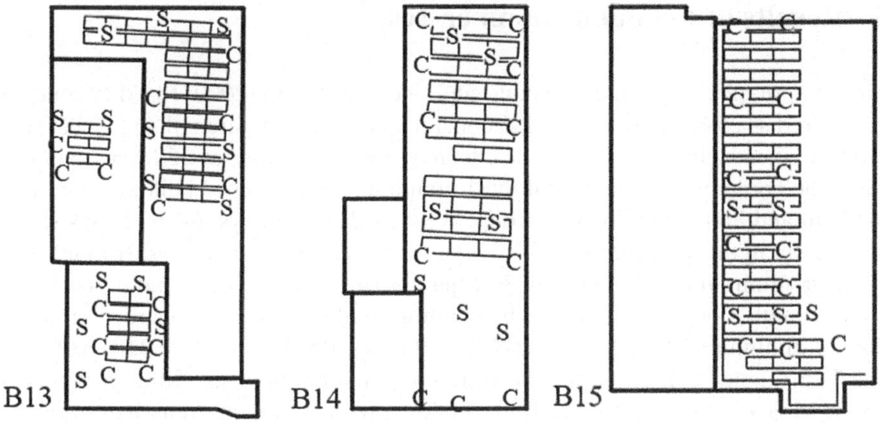

Fig. 5 Roof plan with solar panels showing location of RHT for columns and slabs (C-Columns and S-Slabs)

Generally, a variation of 20 percent to 25 percent is found in the compressive strength of the concrete measured by the RHT. This is because the results of RHT are affected by factors like internal moisture, carbonation of concrete, age of concrete and type of aggregates (Adorjn 2015). More moisture gives less rebound number. In the RHT, when the plunger is pressed against the concrete surface, the spring controlled mass in the hammer rebounds. This give the readings which can be correlated with the graph for the rebound hammer to find the compressive strength of the concrete (Parlewar 2021a, b).

To assess the structural condition of the building the application of RHT was found to be most convenient and faster method to analyse the compressive strength of the concrete old and historic buildings (Parlewar 2021a, b). The rebound hammer was found to be very handy to carry to these building studied in this project. The method to undertake the RHT was as follows: (a) Identification of the location of the structural members for conducting the RHT. These locations were the directly related to the load imposed by the solar panels, (b) cleaning of the surfaces affected by the weather conditions, and (c) Finally, undertaking test by marking nine points in square grid on the surface of the structural member.

The solar panel imposes the load directly on the structural members like slab and beams. Due to the regular geometry, it is difficult to overlay the load on the beams and columns. So, the slabs, takes the load of the solar panel. The executing agencies for the solar plant do not look into the effective distribution of the load on the columns. Hence, the load distribution remains uneven. So, the test was conducted on the top surfaces of the slabs which are directly affected by the weather of the solar panels. Also, main peripheral columns were selected to study the strength of the concrete. These peripheral columns were most affected due to the loading of the solar panels.

5 Results of Rebound Hammer Test

The compressive strength of the building as per the IS 456 (2000) for M25 concrete is required as 25 Mpa. Also, the M25 concrete reaches to the strength upto 25 Mpa in 28 d. So, the original strength of the concrete can be assumed as 25 Mpa. To understand the compressive strength of the 15 buildings, we grouped buildings into low, medium and high strength. The groups were made as follows: (a) buildings showing the strength of minimum 20–24 Mpa in the structural members, (b) buildings showing minimum strength of 25–30 Mpa in structural members, and (c) buildings with strength more than 30 Mpa. The buildings with the strength in structural members of minimum 20–25 Mpa are B1, B2, B4, B5, B6, B7, B8, B9, B10, B11, B13 and B15 (Table 6). The buildings with minimum strength in structural members of 25–30 Mpa are B3 and B14. Lastly, the building with minimum strength of structural members above 30 Mpa is B12. It is found that the structural members in first group are having strength of 20–25 Mpa. This strength is below the strength of the M25 concrete (Figs. 6, 7, 8, 9, 10, 11, 12 and 13). It can be inferred that the buildings in the first group shows lower strength than the required strength of the concrete. As these buildings are more than 40 years old, the structural strength is reduce due to environmental factors. Further, due to the chemical condition in the industrial complex the strength is reduced in the structural members. Once, the solar panels are installed, these members will get directly affected due to the wind pressure and self-load of panels. Following are the inferences based on the RHT: (1) The structural members in the buildings B1, B2, B4, B5, B6, B7, B8, B9, B10, B11, B13 and B15 are showing low strength in the concrete member. Some of the members in these building have strength upto 20 Mpa. The required strength of the concrete members can be assumed for the M25 concrete as 25 Mpa. (2) It can be assumed that the building age is average 40 years old. So, the concrete strength after the 40 years is reduced by 5 Mpa in some of the members. It can be further assumed that during span next 40 years, the building strength will be reduced twice the present reduction in strength. (3) Importantly, installation of solar panels will affect the strength of the structural members. As per the design for module mounted solar panel system, the

Table 6 Table showing results of RHT on 15 Buildings

Building No.	Strength (Mpa)	Building No.	Strength (Mpa)
1	20–68	9	24–66
2	24–62	10	20–66
3	26–68	11	20–74
4	22–64	12	32–62
5	20–45	13	20–64
6	20–47	14	26–60
7	24–70	15	22–60
8	20-47		

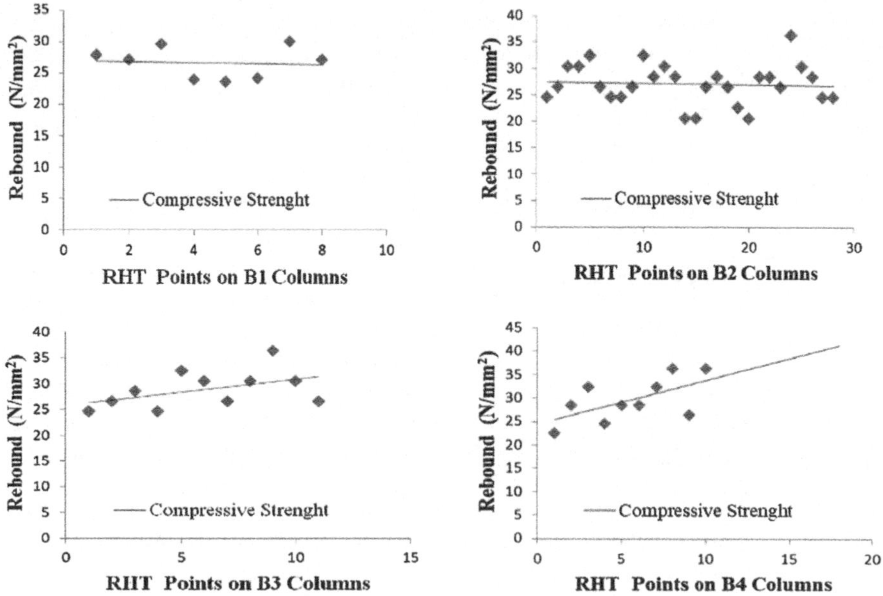

Fig. 6 Results of RHT on columns for building no. B1, B2, B3 and B4

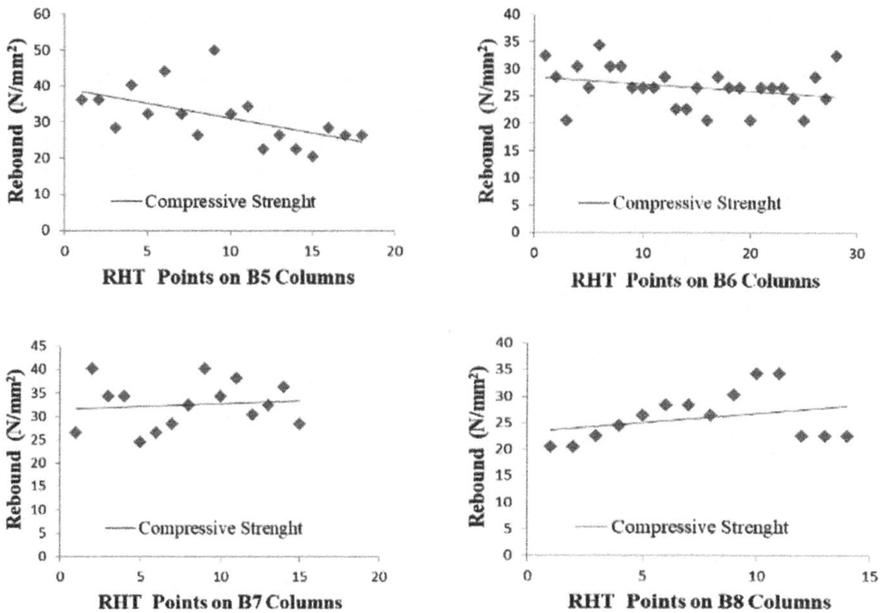

Fig. 7 Results of RHT on columns for building no. B5, B6, B7 and B8

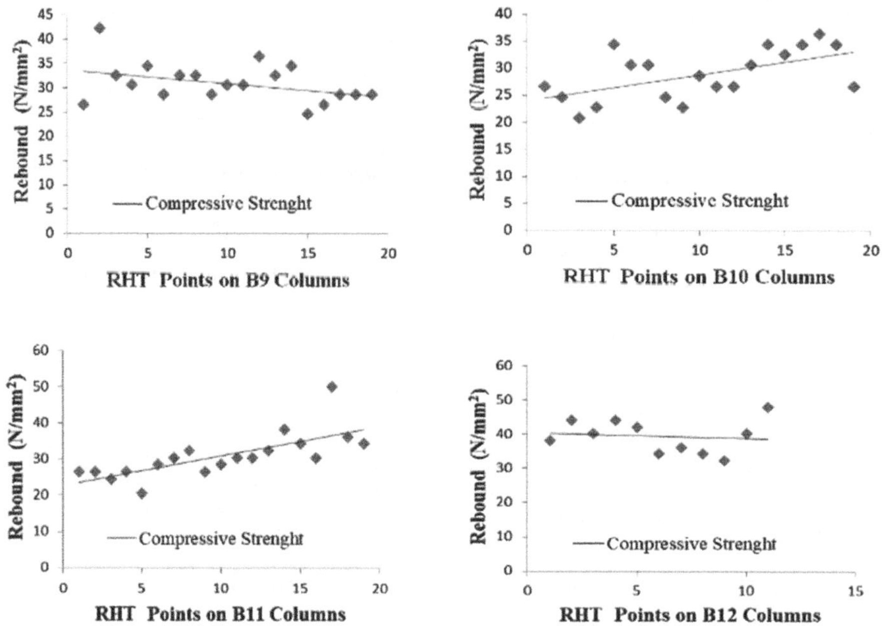

Fig. 8 Results of RHT on columns for building no. B9, B10, B11 and B12

dead load is 0.11 kN/m^2. The wind pressure is 0.8402 kN/m^2 acting on the surfaces of concrete members. Moreover, the each solar panel will also have upward positive force. The uniformly distributed load due to this upward positive force on the structural members will be 0.74 kN/m^2. Finally, the downward negative force due to wind on each module is 0.41 kN/m^2. (4) The acting of these forces on the low strength structural member may further reduce the strength in range of 5–10 Mpa in the span of next 30 years. So, the installation of solar power plant on the low strength members is not feasible due to reduction in the strength. So, it is essential to retrofit this members as early as possible.

6 Application of NDT

To understand the application of NDT for installation of roof top solar power plants, we will first delineate the issue as follows: (1) The wind speed in one of the factor which affect the loading conditions on the roof slab of the solar power plant. Particularly, in Mumbai (Bombay), India the wind speed during the peak month is 14.5 miles per hours. This wind speed engages two types of load on the buildings. One is the upward negative load and other is downward positive load. Hence the NDT test like RHT provides information on the compressive strength of the concrete member

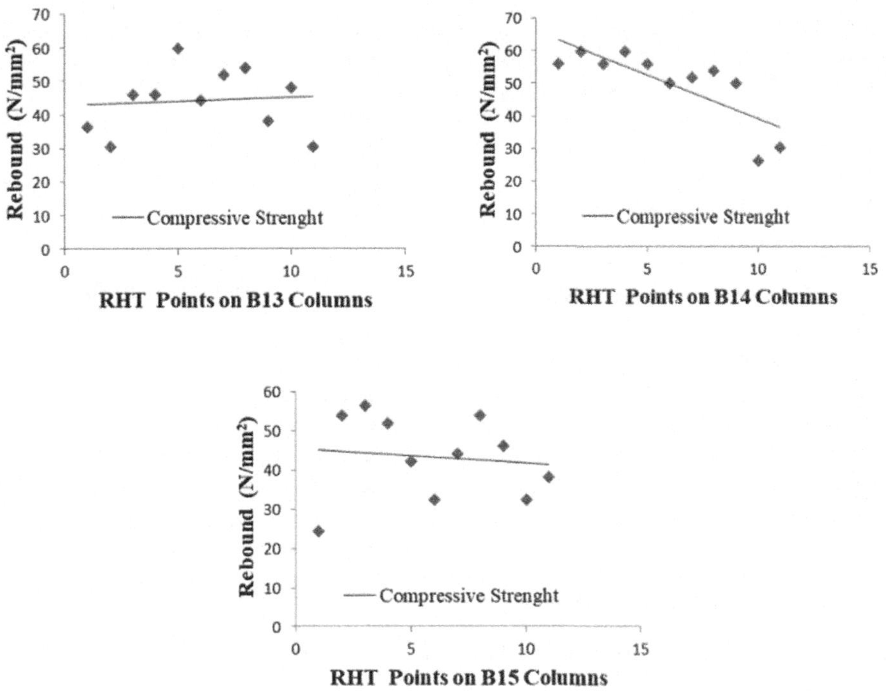

Fig. 9 Results of RHT on columns for building no. B13, B14 and B15

of the building. (2) Significantly, the high wind speed will drastically affect the old deteriorated structural members like beams, columns and slabs. If the member is already deteriorated then it can reach a state of failure during the peak wind speed. (3) Along with the wind speed the structural members are also affected by the dead load of the solar power plant. These loads are weight of the photovoltaic panels, support members, and concrete pedestals. (4) Generally, the life span of the solar panels can be assumed as 30 years. However, it is evident the a solar power plant in a large industrial establishment will last more than 50 years due to good maintenance. Now, we can assume that the structure will also be deteriorating in the passage of time. In case of the project in the petroleum industry in Mumbai, the building are around 30–50 years old. It is expected that the solar power plant will last for further 50 years. So, in the span of next 50 years of a building, it can be assumed that the proposed solar power plant will drastically affect the structural members.

The NDT methods that can be applied for the installation of the roof top solar power plant are:

(1) Rebound Hammer Test: The Rebound Hammer Tests are NDTs that allow the estimation of the concrete quality on-site. The tests uses a Schmidt Sclerometer that measures the superficial hardness of the concrete from the recoil of an incident mass after the collision with the surface being tested. This recoil is

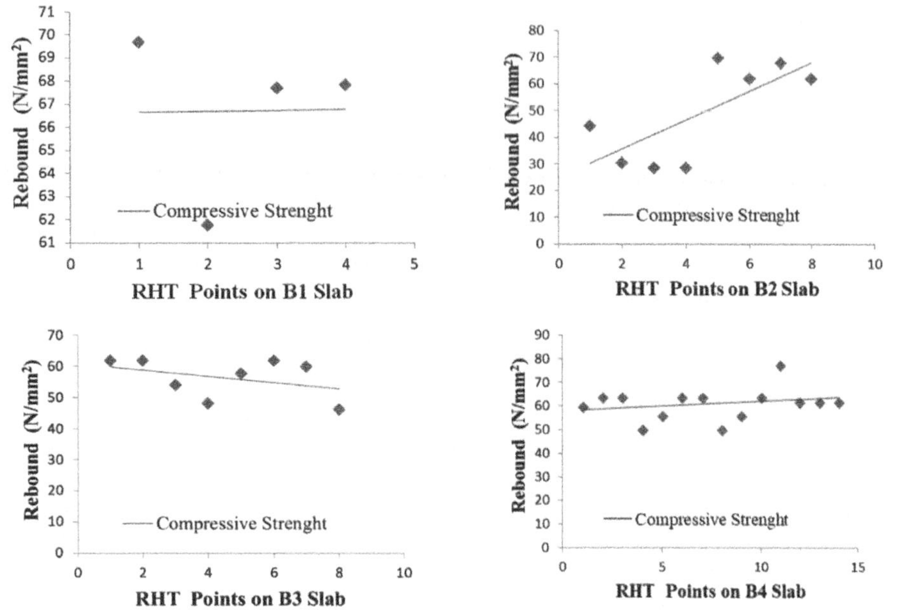

Fig. 10 Results of RHT on slabs for building no. B1, B2, B3 and B4

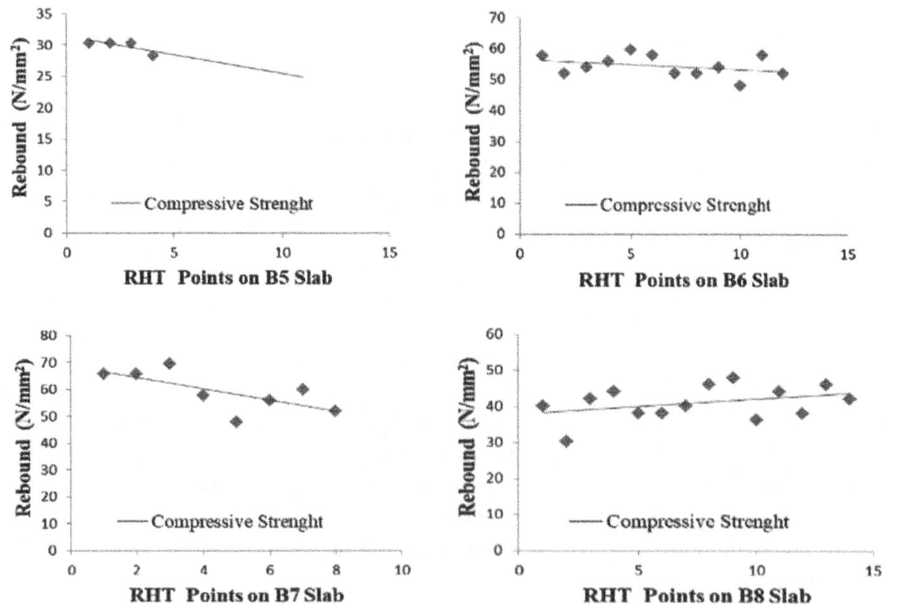

Fig. 11 Results of RHT on slabs for building no. B5, B6, B7 and B8

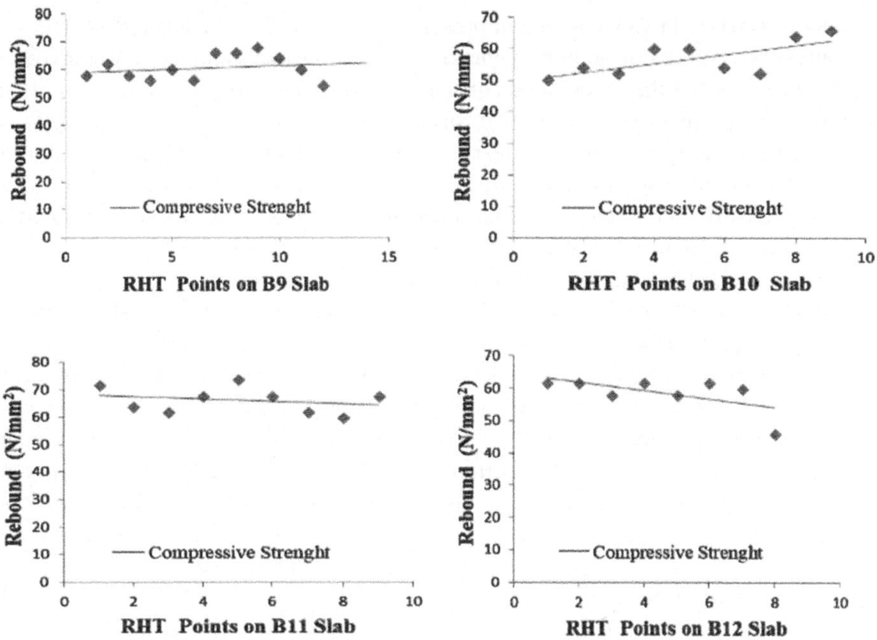

Fig. 12 Results of RHT on slabs for building no. B9, B10, B11 and B12

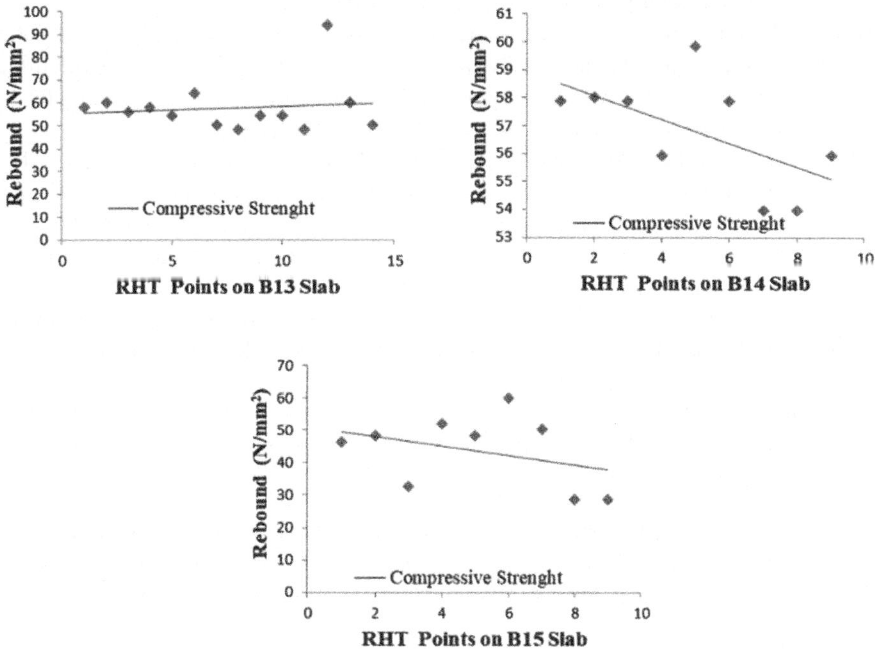

Fig. 13 Results of RHT on slabs for building no. B13, B14 and B15

then converted to a value of compression resistance through an abacus. The test must be carried out on homogeneous concrete surfaces and it usually involves the removal of the carbonated superficial layer by scraping before.

(2) Ultrasonic Test on the Concrete Structures: An ultrasonic pulse velocity test is an in-situ, non-destructive test to check the quality of concrete and natural rocks. In this test, the strength and quality of concrete or rock is assessed by measuring the velocity of an ultrasonic pulse passing through a concrete structure or natural rock formation.

(3) Thermographic Surveys: Thermographic inspection refers to the non-destructive testing of building through the imaging of the thermal patterns at the object's surface. The term thermography alone refers to all thermographic inspection techniques regardless of the physical phenomena used to monitor the thermal changes. The application of a temperature sensitive coating to a surface in order to measure its temperature is a thermographic inspection contact technique based on heat conduction where there is no infra-red sensor involved.

(4) Endoscopic Surveys: An endoscopic survey involves inserting an endoscope camera through an access point or hole to view inside the construction and identify defects. Inserting an endoscope gives the precise vision to identify defects within the construction.

(5) Core Sample of Concrete: The core samples of concrete are cut through the core cutting machine. In this test, three similar diameter and length samples are test in the laboratory for identifying the composition of concrete, compressive strength, carbonation and corrosion in reinforcement.

(6) Load Test: The load test is conducted to by applying the total load of solar power plant on the roof slab by use of sand filled bags. This load is retaining for the period of approximately one week. Then readings are taken at an interval of 24–48 h to underhand the deflection in the structural member.

(7) Flat Jack Test on the Masonry Structures: Flat jack testing method is used to determine the acting stresses (single flat jack) or to evaluate the mechanical parameters of the concrete structures. The testing technique is based on the release of the stress in a small area of a structure by a plane cut perpendicular to its surface.

(8) Cover Test: The cover test estimate the cover over the reinforcement in the structural members. This test is helpful to understand the composition of the structural member.

(9) Foundation Test: Foundation test need to be conducted by opening the floor and remaking it. This test will provide more accurate studies on the stability of the building for imposing any external load.

Thus, various NDT test can be applied on the deteriorated buildings for identifying the structural stability. Moreover, there are various other tests can be applied to the old deteriorated buildings for installation of roof top solar power plants.

Importantly, the application of the NDT for installation of roof top solar power plant can investigate following aspect of the old building: (1) By applying the NDT on columns, beams and slabs, the life span of the structural members can be investigated in the old buildings. (2) It is well evident that the structural members will further deteriorate due to weather and installation of solar power plant. Thus, we can investigate and plan for the retrofitting of the structural members of the building. (3) This investigation can easily identify the members to be strengthened to prevent failure by installation of roof top solar power plant. (4) This application can also predict the cost effective retrofitting of the structural members in a building. (5) Moreover, we can identify the best possible location for uniformly distribution of load of concrete pedestals, steel framing and solar panels.

7 Conclusions

The RHT on the 15 buildings for installation of solar power plant has shown that the compressive stregth of the concrete had reduced in the structural members. Following are the inferences of application of NDT for installation of solar power plant: (1) The RHT for 15 buildings had inferred that majority of building are having low compressive strength. So, a retrofitting was proposed for these building. (2) Based on the proposed retrofitting, the buildings were recommended suitable for installation of module mounted solar panels. (3) The module mounted solar panels will have dead load of kN/m^2. It will have wind force of 0.74 kN/m^2 in upward negative detection and 0.41 kN/m^2 in downward positive directions. (4) Thus, the applications of NDT for installation of roof top solar power plant investigate the requirements for retrofitting for buildings for long term stability. This application can be undertaken through various types of tests like RHT, UPV test, load test, core sample test or others. The roof top solar power plant have many advantages for individual house and also to large industries. The solar power generation produces green energy that can be directly utilized for various purposes. Further, similar research can be carried out on the old buildings that are deteriorated in passage of time for installation of rooftop solar power plants.

References

Borosnyi Adorjn (2015) NDT assessment of existing concrete structures: spatial analysis of rebound hammer results recorded in-situ. Eng Struct Technol 7(1):1–12. https://doi.org/10.3846/2029882X.2015.1085331

Katalin S, Adorjan B, Istvan Z (2015) Understanding the rebound surface hardness of concrete. Civil Eng Manag 21(2):185–192

Parlewar P (2021) Rehabilitation of K. G. Road Complex at New Delhi, India. In: Delgado J (eds) case studies in building rehabilitation. Building Pathology and Rehabilitation, vol 13. Springer, Cham

Parlewar P (2021) Rehabilitation of historic chancery building, Yangon, Myanmar. In: Delgado J (eds) case studies in building rehabilitation. Building Pathology and Rehabilitation, vol 13. Springer, Cham

Sauca A, Milchi T, Gobesz F (2019) Wind loading on solar panels. Mszaki Tudomnyos Kzlemnyek 10(1):73–78. https://doi.org/10.33894/mtk-2019.10.10

Local Reinforcing of RC Columns by Steelwork: A Parametric Study

Antonio Formisano

Abstract In the work the problem of local reinforcement of columns belonging to existing reinforced concrete (RC) buildings by means of steel angles and batten plates is examined. Initially, a sophisticated numerical FEM model has been calibrated through the calculation code ABAQUS on the basis of a laboratory experimental test on a RC column reinforced with the aforementioned metallic elements and subjected to both vertical loads and earthquakes actions. Subsequently, an extensive parametric analysis on the above reinforced columns has been conducted by varying the main geometric parameters characterising the reinforcement system, such as the angle height and the width and thickness of batten plates. Numerous configurations of reinforced columns have been inspected by means of static non-linear analyses, always conducted with the ABAQUS program, in order to identify a series of optimal reinforcement solutions capable of combining the best structural performances with the lowest possible cost. Finally, the selected optimal configurations have been compared to each other by performing cyclic analysis to identify the reinforcement system providing the best seismic response in terms of dissipated energy.

Keywords FEM model · Reinforced concrete columns · Numerical methods

1 Introduction

The worldwide magnitude of degradation of reinforced concrete structures is increasing at alarming rate. The causes are several, including concrete decay, transformation of the use class and loading requirements, corrosion of reinforcements, as well as inaccuracies during design and construction (Raza et al. 2019). An additional factor is presented by the risk of a seismic event, which highlights the need of increasing the performance of buildings (Paulay and Priestley 1992).

A. Formisano (✉)
Department of Structures for Engineering and Architecture, University of Naples "Federico II", Piazzale Tecchio 80, 80125 Naples, Italy
e-mail: antoform@unina.it

J. M. P. Q. Delgado (ed.), *Durability of Concrete Structures*, Building Pathology and Rehabilitation 16, https://doi.org/10.1007/978-3-030-62825-3_2

Requalification of the urban order represents the actual priority for many countries, such as Italy, that base their politics on low land consumption and adversity in demolishing old buildings. In Italy, the new real estate cycle, which started in 2008, discourages interventions on the outside of buildings, thereby endorsing urban requalification, redesign of existing interior spaces and change of the use class (Ministry of Infrastructure and Transport 2008). Strengthening of existing buildings is certainly more sustainable considering both the preservation of resources, such as materials and time, and the reduction of the ecological impact of the construction business (Raza et al. 2019).

In this context, the concepts of repair and retrofitting assume fundamental importance in the modern engineering practices. Repair refers to the modification of a structure after damage occurs, whereas rehabilitation before damage is referred to as retrofitting (Wu et al. 2006).

In line with a global trend and to avoid post-earthquake disasters, it is clear how retrofitting is gaining more and more attention both in terms of techniques, but also of innovative and renewable resources.

A large portion of the Italian structures made of masonry or reinforced concrete was designed and built prior the 1970s. Existing buildings were frequently designed considering vertical loads only, contrary to the current safety regulations, that are based on exceeding performance levels (Performance Based Design). This is confirmed by the rules for the execution of works R.D.2229/1939, which for many years regulated the design of reinforced concrete and masonry constructions. It stated that "potential dynamic actions are taken into account by increasing the loads in relation to their nature and structural typology". All this highlights that the design did not consider any construction detail aimed at facing exceptional events, such as earthquakes. Instead, the existing buildings were designed so that the strengths of the individual structural elements were greater than the stresses induced by the action loads.

During the 1970s, the first guidelines introduced the design based on the structure dissipative capacity through application of the so-called hierarchy design criteria. However, only recently (starting from OPCM 3274/2003), these were implemented by Italian national legislation. In this framework, in order to adequately deal with an earthquake, it is necessary to provide the structure with sufficient resistance, but also with adequate ductility. The latter represents the structure ability to withstand significant deformations beyond the elastic limit and before reaching failure. It constitutes a fundamental mechanical property not less important than the resistance itself.

With reference to reinforced concrete (RC) structures, their optimal seismic behaviour is linked to the formation of global collapse mechanisms, that involve the entire structure, rather than local ones. Most of the involved seismic energy is dissipated through the formation of "plastic hinges" at the ends of elements according to the resistance hierarchy principle, thereby avoiding local collapses in critical fragile areas. Considering all the above, the existing RC buildings are characterized by: (1) low amount of longitudinal reinforcement in the columns compared to shear reinforcements, (2) beams with a an abundant longitudinal and transversal reinforcement compared to the columns. These peculiarities identify a "strong beam-weak column" collapse mechanism (Aboutaha et al. 1999; Alcocer 1993; Alcocer and Jirsa 1993; Valluvan et al. 1993).

For what concerns RC columns with deficient shear reinforcement (Seible et al. 1997), the possible failure mechanisms are three:

- Shear failure;
- Collapse of flexural plastic hinges because of consistent degradation dependent on: (1) flexural cracking, (2) concrete spalling, (3) reinforcement buckling;
- At the bottom of the column, where the bending moment reaches its maximum, the potentially occurring plastic hinges can lead to lap slices in the longitudinal reinforcement.

Lynn et al. experimentally investigated the vulnerability of such columns (Lyn et al. 1996).

In the scientific literature there are different local strengthening techniques for reinforced concrete columns, namely reinforced concrete jacketing, composite materials based on Fibre-Reinforced Polymers (FRP) and steel jacketing. With reference to reinforced concrete columns, steel jacketing based on the use of steel angles and batten plates is one of the most common strengthening technique in Europe. With this kind of strengthening technique, it is essential to guarantee the correct load transmission from the beam to the strengthened column. This chapter wants to analyse the reinforcement technique of a RC column strengthened with corners steel angles joined by means of steel welded batten plates. After a literature review on this topic, an experimental campaign coming from literature has been considered as the basic reference test. This experimentation has been chosen because it deals with a full-scale RC column, having height, cross-section size and reinforcements like those of columns belonging to buildings designed without reference to anti-seismic criteria. The mentioned experimental test has been numerically simulated through the well-known ABAQUS/CAE finite element analysis program. The FEM model of the strengthened RC column has been subjected, according to the experimental evidences, to both gravity and seismic loads, the latter applied under form of either monotonic actions or cyclic loads. Different configurations of the reinforced column, by changing the main geometric features and mechanical parameters of the steel reinforcing system, have been numerically tested to individuate the best reinforcement system from the structural and economic viewpoints.

2 Design Interventions on Existing Structures

The current technical standards for constructions (Ministerial Decree 17/01/2018) classifies the anti-seismic interventions into three types:

- Repair or local interventions: interventions that affect individual structural elements. They do not reduce the pre-existing safety conditions in any case and do not change the overall behaviour of the structure;
- Upgrading interventions: interventions aimed at increasing the existing structural safety, without necessarily reaching the safety levels indicated by the regulations for new constructions. These changes affect the overall behaviour of the structure;
- Retrofitting interventions: interventions aimed at increasing the existing structural safety by reaching the safety levels set for new buildings. These actions alter the overall behaviour of the structure.

In this framework, it should be noted that, despite being operations on single elements, local interventions, if adequately calibrated, can also be considered as global interventions, because they influence the behaviour of the entire structure. Therefore, it is always fundamental to examine the influence of various local interventions as a whole on the construction (Ghimire 2019).

Choosing the elements to be retrofitted and designing the intervention can be time-consuming, especially on buildings made of irregular structures with several columns. Approaches based on non-linear static (pushover) analysis to assess the pre- and post-intervention status of the building require a great number of attempts. In order to balance the effectiveness of the intervention, the costs and the time for analyses, several structural optimization algorithms have been suggested. Most of such techniques were developed for the design of new buildings made of RC or steel structures. Among these, Optimality Criteria approaches (Zou et al. 2007), Genetic Algorithms (Liu et al. 2003; Greco and Marano 2011), Discrete Evolution Strategies algorithm (Lavan and Dargush 2009; Papavasileiou and Charmpis 2016) and Meta-Heuristic Search techniques, such as the Harmony Search algorithm (Akin and Saka 2015), are to be listed. Structural optimization problems are generally easier to solve for steel, but more complicated for RC due to the complexity of reinforcement modelling. The above-mentioned studies on RC structures are focused on multi-storey frames but are applicable to regular structures only. Recently, an investigation study was conducted on 3D RC building structures with regular and irregular plan shapes using numerical tool for the quantitative estimation of the life-cycle cost and vulnerability assessment (Mitropoulou et al. 2011).

With the new trend of artificial intelligence, methodologies aimed at optimizing strengthening and retrofitting are currently being developed as a matter of strong interest. Many retrofitting devices were analysed: carbon fibre reinforced concrete slabs (Chaves and Cunha 2014), FRP jackets (Chisari and Bedon 2016), dissipative bracings (Braga et al. 2019) and steel jacketing (Di Trapani et al. 2020).

Among the local interventions on single elements, altering the stability of the column was believed a poor result, since it influences the dynamic properties of the structure. However, it has recently been highlighted that such an alteration can result in an overall positive effect (Raza et al. 2019).

The first hypothesized retrofitting intervention was a supplementary layer of concrete with longitudinal and transversal reinforcement over the RC column affected by one of the previously mentioned failure mechanisms. The implementation is presented by the use of steel instead of reinforced concrete (Frangou et al. 1995; Ramirez 1996; Dritsos 1997). Local interventions aim at enlarging the ductility of the structure with the minimum increase in strength. In this way, the variation of demand in terms of displacement remains unchanged but an increased capacity is obtained (Ghobarah et al. 1996). Concrete jacketing, instead, often increase the stiffness due to the typical increase of the column cross-section.

Concrete jacketing was proven to be an effective repair mechanism after the Mexico City earthquake of 1985, yet labour-intensive (Rodriguez and Park 1991) especially because of the need of temporary suspension of the building use during the application. Overall, concrete jacketing is referred-to as global technique when the longitudinal reinforcement of the jacket goes through the slab and the beam-column joints are covered by new concrete.

Among the possible techniques for interventions, the use of steel is particularly effective and in line with actual environmental criteria. Steel stands out for its intrinsic characteristics, such as lightness, ductility, prefabrication with reduced operations on the construction site, high strength-to-weight ratio and excellent durability even in extremely corrosive or saline contexts, if properly treated. Among several techniques that are currently available, steel jacketing is the most common intervention (Tamai et al. 2000). It consists of plates, strips and angles that can be composed with several arrangements, including steel caging, steel encasement, steel plates with anchors and circular/oval shaped steel jacket. All these are reviewed in detail in (Wu et al. 2006).

In comparison to concrete, steel jacketing provides many advantages, such as a minor increase of the cross-section, faster and straightforward construction, no use interruption, cheaper interventions and overall inferior increase of the additional column stiffness (Tamai et al. 2000).

Compressive longitudinal force on RC columns leads to a possible lateral expansion. To limit such an effect, transverse reinforcement is employed for a confinement action opposing the concrete enlargement. As pointed out by Braga et al. (Braga and Gigliotti 2006), the confinement response supplied by the transversal confinement can be referred-to as "passive confinement". This relays on the observation that the steel intervention effectively withstands concrete expansion only for high strain levels in concrete, that is when the force almost equals its uniaxial compression strength.

In particular, steel angles connected by welded plates (Frangou et al. 1995) have been employed for interventions on columns to increase their load-bearing capacity (lateral confinement effect) and shear strength, thus minimizing their lateral deformations (Nunziata 2011). The confinement action on the cross-section is to be considered summable to that of the brackets, therefore the performance levels vary along the longitudinal development of the column. However, a model with a single concrete stress-strain law can be used, as the action of the steel jacketing prevails. This is valid under the hypothesis that the confinement exerted by the transverse reinforcement takes place in plane deformation conditions (Braga and Laterza 1998).

Several authors provided indications for determining the constitutive laws of concrete for the effectively confined concrete and for the unconfined concrete (Razvi and Saatcioglu 1992; Montuori and Piluso 2009).

Two main methodologies for steel jacketing of columns are being employed:

- Only the column is reinforced with the additional steel elements ending before its extremitics. This practice is extremely common because of the difficulty in setting up a moment resisting connection to the beam/slab. The benefit relays on a strong concrete confinement, that leads to expanded deformation capacity. Eurocode 4 (CEN—European Committee for Standardization 2005a, b) and the Italian Technical Codes for Constructions (Ministry of Infrastructure and Transport 2018) assess the expansion in bearing capacity based only on the confinement effects. In case of ideal end connections, Eurocode 4 agrees with the consideration of angle contribution.
- The intervention is designed to ensure a connection between the jacket and the horizontal elements of the pre-existing structure. Moment resisting connections are designed to increase the section flexure resistance.

Despite planning an intervention of the first type, the section will always experience a remarkable increase of flexural strength because of the friction forces between the steel angles and the concrete column (Campione et al. 2017).

All these remarks show the fundamental importance of the connection between the angles and the structure. The increase in longitudinal steel reinforcement induces an increased flexural strength and ductility of columns towards compressive-bending stresses, which depends on the connection degree between the angles and the structure. Therefore, it is fundamental to properly anchor the steel angles to the pre-existing elements to ensure robustness. Particular attention should be paid to the prior removal of deteriorated concrete and restoration of the core (Lippi et al. 2018).

The transverse plates connect the longitudinal angles by means of corner welds (Dritsos and Pilakoutas 1994; Georgopoulos and Dritsos 1994). The former are heated before welding, so that the contraction during cooling leads to a proper contact between steel and concrete for a correct loading of the system.

Besides being particularly effective for barycentric loading, as proved by its broad use in Japan, Taiwan and the United States (Rodriguez and Park 1991; Li et al. 2005), steel jacketing was proven to improve the carrying capacity of columns that withstand eccentric loading (Ezz-Eldeen 2016) and was largely employed for retrofitting of RC columns in the aftermath of a seismic event (Murty et al. 2005; Wasti and Ozcebe 2006).

3 Literature and Standard Provisions on Design Rules

In recent years, many studies on RC columns retrofitted using steel jackets (Zhong et al. 1986; Unjoh and Kawashima 1992; Priestley et al. 1994; Aboutaha et al. 1994) aimed at providing analytical methods for design. Dritsos and Pilakoutas (Dritsos et al. 1993) proposed a procedure to quantify the confinement of concrete due to the steel cage assuming that the combined response of steel and concrete at the interface is propelled by Poisson's expansion and interface friction. Masri and Goel (Masri and Goel 1996) developed a method to determine the jacketing bending contribution taking into account the observed load distribution between the angles and battens. The latter was based on laboratory tests on a 1/3 scale two-storey building reinforced with steel bracings and jacketed columns. The analytical method underestimated the observed strength increases, because it was not able to calculate the benefit of jacketing on the column core. Nagaprasad et al. (Nagaprasad et al. 2009) carried out an experimental campaign on RC columns before and after reinforcement by steelwork aimed at increasing both their axial and cyclic lateral load. The authors provided a method capable of considering the confinement effect of steelwork on the existing column and anticipating failure of the steel cage.

The section C8.7.4.2.2 of the explanatory circular of NTC 2108 (Ministry of Infrastructure and Transport 2018) provided the mathematical expressions to evaluate the increase of both axial and shear strengths and deformation, as well as the confinement action, due to steel reinforcement.

The additional shear strength offered by the jacket, provided that the stress does not exceed 50% of the yield value, is given by the following equation:

$$V_j = 0,5 \cdot \frac{2t_j}{s} \cdot b \cdot f_{yw} \cdot 0,9 \cdot d \cdot \text{cotg}\,\theta \tag{1}$$

where t_j, b and s are thickness, width and spacing of steel battens ($b/s = 1$ for continuous jacketing), d is the cross-section height, f_{yw} is the steel yield strength and θ is the inclination of shear cracks.

The confinement effect of the steel jacket is assessed, as for the stirrups, with reference to the geometric percentage of reinforcement present in each of the transverse directions, with the following expression:

$$f_{cc} = f_c\left[1 + 3,7\left(\frac{0,5\alpha_n\alpha_s\rho_s f_y}{f_c}\right)^{0,86}\right] \tag{2}$$

where: ρ_s is the transverse reinforcement volumetric ratio, equal to $2(b+h)t_s/(bh)$ for continuous jacketing and $2A_s(b+h)/(bhs)$ for discontinuous jacketing, being b and h the width and height of the column section, respectively, s the spacing of battens and t_s the thickness of the steel jacketing; α_n and α_n are the confinement efficiency factors in the section and along the element, respectively, given by:

$$\alpha_n = 1 - \frac{(b-2R)^2 + (h-2R)^2}{3bh} \tag{3}$$

$$\alpha_s = \left(1 - \frac{s-hs}{2b}\right)\left(1 - \frac{s-hs}{2h}\right) \tag{4}$$

where R is the rounding radius (if any) of the section edges (for angles, R can be assumed equal to the lesser between the angle side length and 5 times the angle thickness), b and h are the cross-section dimensions and h_s is the height of the discontinuous bands (if the jacket is continuous, $h_s = s$).

The ultimate deformation of the confined concrete is provided by the following expression:

$$\varepsilon_{cu} = 0,0035 + 0,5 \cdot \frac{0,5\alpha_n \alpha_s \rho_s f_y}{f_{cc}} \tag{5}$$

Steel jacketing can provide an effective clamping action by adherence in the junction areas. To this end, it is necessary to respect the following indications: (1) the jacket should be extended beyond the overlying region for a length representing at least 50% of the overlap, (2) the jacket must be fixed on the structural element under pressure with two rows of high strength bolts (Aboutaha et al. 1996).

Generally, the metal jacket must remain in the elastic field so that its contribution is added to that of the unreinforced column. This issue is essential to limit the width of potential cracks in the concrete, thereby guaranteeing its integrity and preventing the spalling of the concrete covering (Sakino and Sun 2000).

If the intervention involves columns over several levels, it is possible to restore their continuity by means of bars passing in the concrete beam-column node and welded at their ends to the corner profiles. A similar solution can be adopted for foundation when acting on ground floor columns.

4 The Reference Experimental Test

The reference experimental campaign was performed in Spain by Adam et al. (Adam et al. 2007) on RC columns strengthened with steel angles and batten plates. The column had a cross-section of 30 × 30 cm with height of 250 cm. It was longitudinally reinforced with 4Φ12 bars and Φ6 stirrups having constant spacing of 20 cm. 27 × 16 × 0.8 cm batten plates, 41.5 cm spaced each to other, and 25 × 8 × 0.8 cm angles in each column corner were used as reinforcing elements.

The column was made of a medium-low quality concrete with a compression stress f_c between 10.6 MPa and 15.5 MPa. The steel used for angles was of S235 type, while the mortar used to connect the angles to the concrete had an average Young modulus E_m equal to 25000 MPa with a Poisson's ratio of 0.2. Two types of specimens were fabricated and tested: SPECIMEN A, having angles and batten plates throughout the whole length of the element, and SPECIMEN B, with steel L-shaped members close to the column head. The test was based on the application of a vertical load, through a hydraulic press with maximum capacity of 5000 kN, able to apply displacements of 0.5 mm/min. Fourteen displacement transducers were installed on the steel elements, whereas the concrete parts have been monitored in eight positions. Some views of the experimental specimens are depicted in Fig. 1.

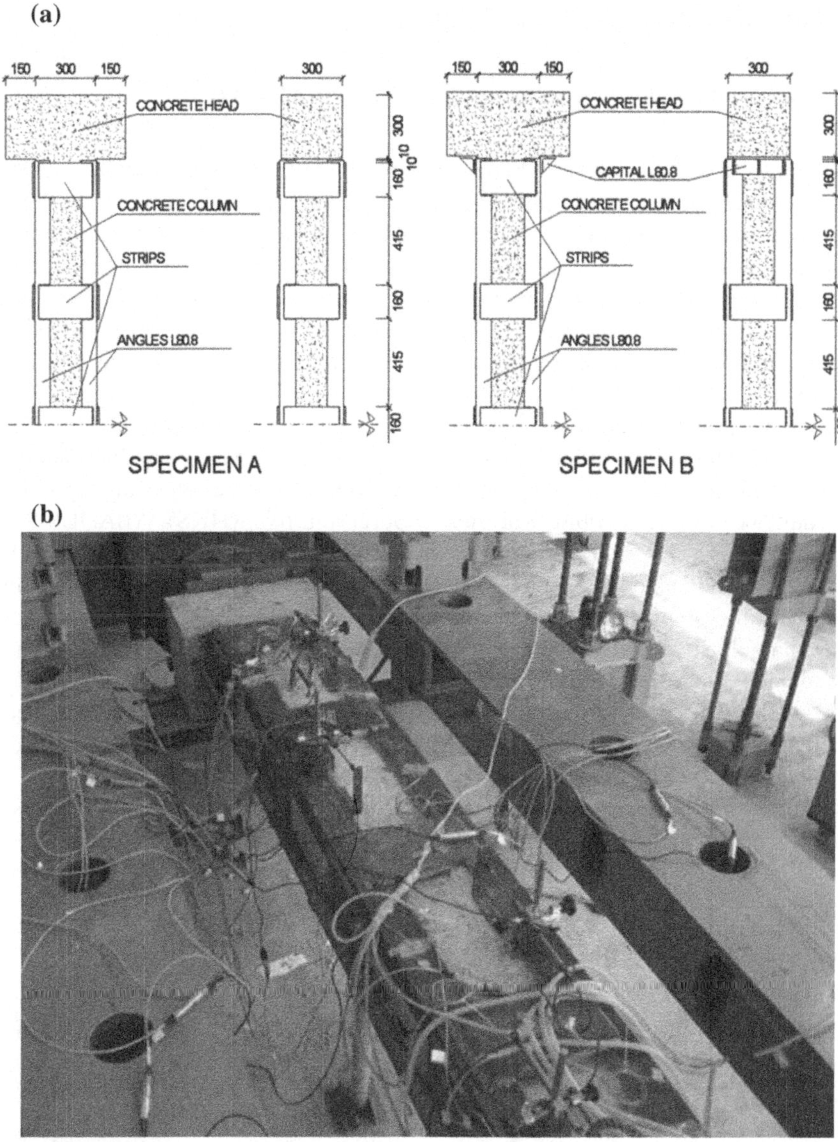

Fig. 1 The tested columns (**a**) and one of the specimens ready for test (**b**) (Adam et al. 2007)

5 The ABAQUS/CAE FEM Model

5.1 Numerical Calibration of Experimental Tests

Based on the experimental tests before illustrated, the geometric modelling of both the RC column and the steel reinforcement system has been performed by using the *Part Command* from the ABAQUS menu. In Fig. 2a the FEM model of the column head section has been shown. For mechanical characterisation of the strengthened column, three different materials, namely concrete, reinforcement steel and steel, have been defined through assignment of densities, elastic properties and plastic features according to the data deriving from the experimental campaign illustrated in the Sect. 4.

After the geometrical model has been setup and material properties have been defined, the interactions properties among structural parts have been considered. In particular, identification and assignment of appropriate contact types among system parts have been done (Hibbitt, Karlsson & Sorensen, Inc., (HKS), ABAQUS Standard/explicit user's manual, Version 6.7, USA 2008). The contact model chosen has always been the standard Surface-to-Surface contact. The contacts distinguishing the structural systems interacting each to other are framed within two different categories, namely steel-concrete and steel-steel contact interactions (Fig. 2b, c). These two contact categories are marked by a friction coefficient equal to 0.3 and 0.4, respectively. A total of forty-eight interactions, namely forty related to the steel—steel contact type and eight concerned the steel—concrete one, have been defined in the implemented FEM model.

On the other hand, tie constraints (Fig. 2d) have been used to model interactions between steel angles and batten plates, that are joined by welding connections, preventing relative movements.

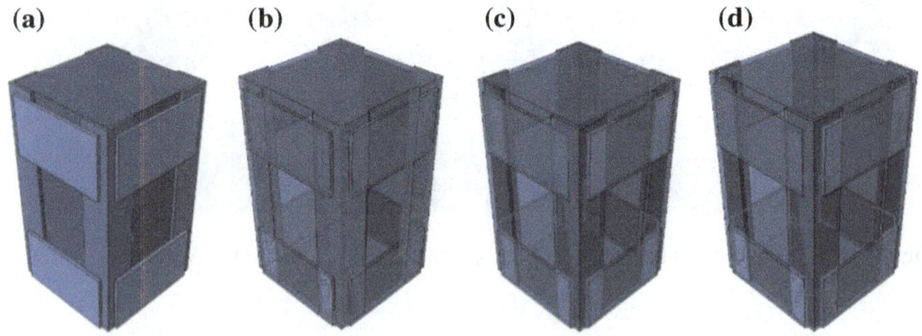

Fig. 2 Column head section: complete pattern (**a**), steel-concrete interaction (**b**), steel-steel interaction (**c**) and tie constraints between angles and batten plates (**d**)

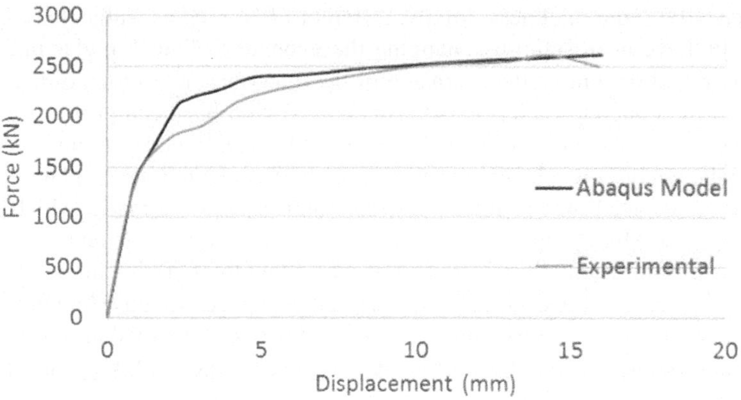

Fig. 3 Comparison between the experimental curve and the ABAQUS one for the reinforced column

Regarding the load application, two analysis steps have been considered: an initial step, where the dead load has been applied, and a second step, where the application of vertical displacements occurs.

The selected investigation procedure is an implicit dynamic analysis. This analysis consists of a direct dynamic integration, used to investigate the dynamic, linear or non-linear response of the examined system. This analysis type has been chosen to correctly calibrate the experimental results considering the strong deformations of angles and batten plates, as well as the large number of interactions and constraints, which did not allow to attain the convergence of results.

The model mesh has been made by choosing continuous three-dimensional first order elements defined by eight nodes with linear integration. To reproduce faithfully the monotonic tests, a displacement control analysis has been performed.

From the comparison between the experimental test and the numerical simulation obtained by means of the ABAQUS program (Fig. 3), a perfect correspondence of results in terms of initial stiffness and maximum strength attained is noticed. Contrary, in the numerical analysis the beginning of the plastic field is characterised by strengths slightly greater than those achieved from the experimental test. This is due to the higher imperfections affecting the structural system in the reality.

5.2 *The Parametric Analysis*

The main objective of the research activity is to identify, through a parametric analysis, which materials and geometrical parameters of the steel reinforcing system are able to achieve the best increase of the structure performances in terms of strength and stiffness. It is worth to be noticed that, as basic assumption, angles and batten plates have the same thickness. Firstly, different FEM models, subjected to gravity loads only, have been setup by changing the geometry of steel angles in terms of both length and thickness, the extension of the intervention in terms of used batten plates and the concrete type (Fig. 4). This comparison has been performed for the attainment of the same ultimate displacement of the reinforced column.

From the analysis results it appears that the best angle and batten plate thickness is 8 mm, since it allows to attain the maximum system strength, which is only slightly exceeded by 10 mm thick elements. Moreover, about angle length, an optimal dimension has not been individuated. However, a length of 80 mm allows to attain a substantial increase of resistance in comparison to angles with base length of 60 and 70 mm. In addition, the performance comparison between the total jacketing and the use of batten plates with two different spacing (127.5 mm and 415 mm) has been done.

It is noticed that the best system is that with 415 mm spaced batten plates, since the use of a pitch of 127.5 mm does not allow to significantly improve the system performance. Finally, when the system tested in (Adam et al. 2007) is used, the concrete class modifies significantly the system strength and the best increase of performance is achieved with the C20 concrete type.

After the analyses under vertical loads, the combination of gravity actions and seismic forces on the reinforced column has been applied (Fig. 5). Each column model has been solicited by two steps: in the first step the vertical action is simulated, while in the second one the seismic action is applied with a displacement increase law of 0.1 mm/s up to a maximum lateral displacement of 100 mm. The *Amplitude command* in ABAQUS has been defined for defining this analysis step.

The setup of the above twenty-two FEM models subjected to horizontal and vertical loads has shown relevant results. In particular, seven parameters have been identified as representative for the problem characterization: batten plate height and thickness, angles length and thickness, concrete class, members steel type and bars and stirrups steel class. The first four parameters are those considered as variable factors to obtain the choice ranges for the optimum reinforcement solution. Contrary, the last three parameters are considered as constant factors, since their variability did not modify significantly the experimental results. For this reason, a concrete with f_{ck} 25 MPa, a S275 carpentry steel and a Fe32k steel type for bars and stirrups have been assumed as basic materials for FEM models.

Based on these hypotheses, fifteen numerical models have been generated. They have been divided into three groups of five models and for each group three parameters (height of batten plates h, length of the angles L and thickness of batten plates and angles t) have been changed one by one, by leaving fixed the other two. In particular, for each model, there is only one variable parameter, whereas the others are fixed and defined as follows: $h = 160$ mm, $L = 80$ mm and $t = 8$ mm. Results of this parametric analysis are shown in Fig. 6a, b.

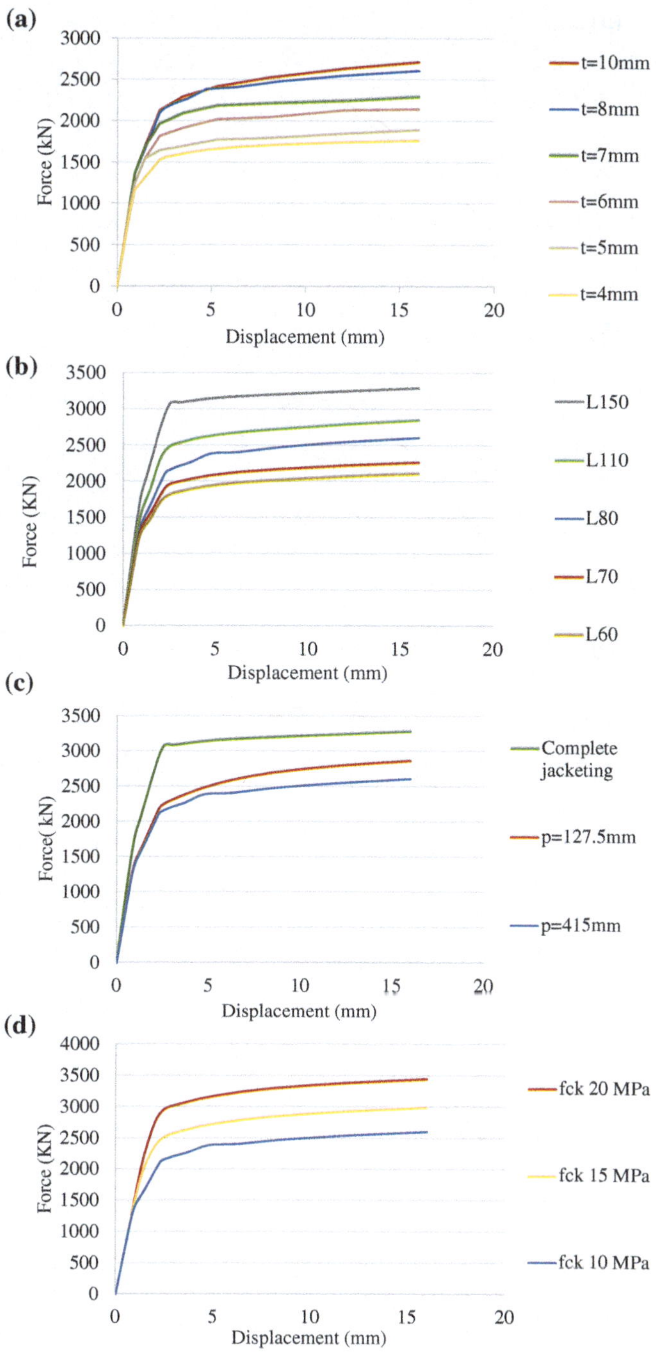

Fig. 4 Numerical analyses results: variation influence of angle thickness (**a**), angle length (**b**), number of batten plates used (**c**) and concrete class (**d**)

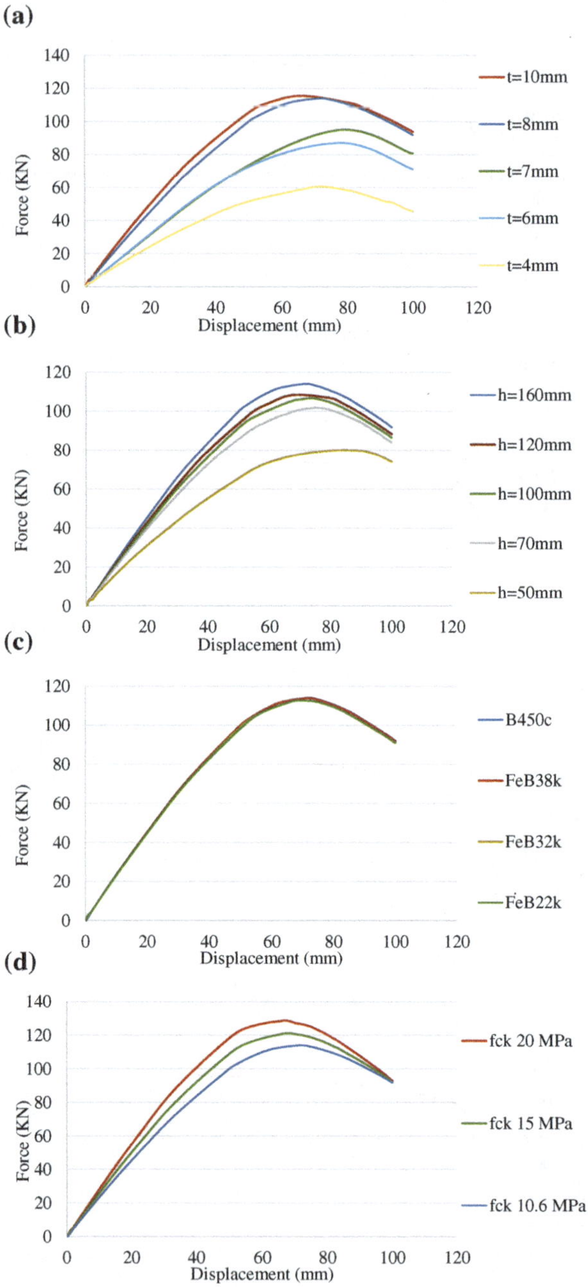

Fig. 5 Analysis under vertical loads and seismic actions: influence of steel profiles thickness (**a**), batten plates height (**b**), steels of bars (**c**) and concrete type (**d**)

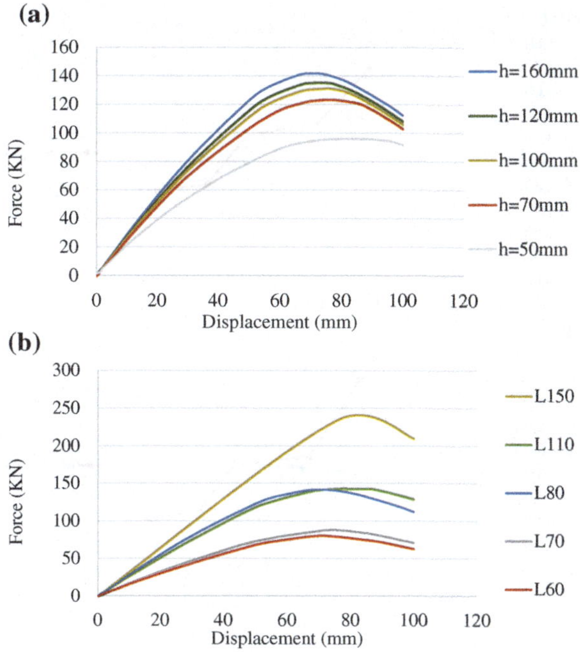

Fig. 6 Influence of the variability of height of batten plates *h* (**a**) and length of angles *L* (**b**) on the behaviour of examined strengthened columns

The fifteen numerical models allow for finding the best strengthening configuration depending on the ultimate resistance required. To this purpose, the results have been plotted in diagrams where the ultimate axial load $N_{u,r}$ and bending moment $M_{u,r}$ of the strengthened column with variable parameters of the reinforcement system, normalised with respect to the ultimate resistances of the original column without reinforcements ($N_u = 1112$ kN and $M_u = 64.2$ kNm), have been represented versus the *h/L* ratio, that is the ratio between the batten plates height and the angles length. In the two diagrams of Fig. 7a, b, the intersection points among curves individuate the two best reinforcement solutions, namely the 125 × 115 × 8 mm and 80 × 75 × 8 mm configurations, where the first, the second and the third number are batten plates height, angle length and batten plates and angles thickness, respectively.

Cyclic analyses under horizontal loads have been performed on these reinforced columns to evaluate the increase of dissipated energy with respect to that offered by the reference column strengthened with the reinforcing system tested in (Adam et al. 2007). In Fig. 8 the results of these tests on the two retrofitted columns under four loading cycles have been reported.

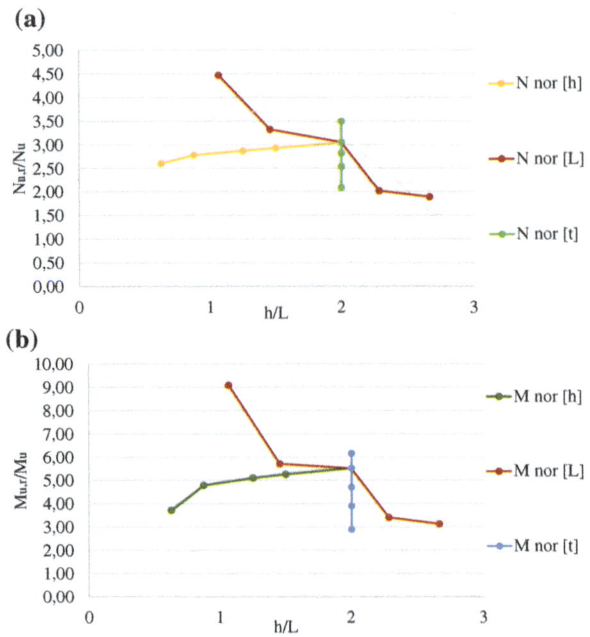

Fig. 7 Individuation of best local reinforcing solutions in terms of compression axial load (**c**) and bending moment (**d**)

Cyclical analysis has denied the static analysis results. In fact, the 125 × 115 × 8 mm configuration is better than the 160 × 80 × 8 mm one. This result is also supported by economic analysis, considering that the considerable increase of dissipated energy compensates the marginally higher cost of the used reinforcement system (Fig. 9).

6 Concluding Remarks

The large computational campaign performed on RC columns reinforced with steel angles and batten plates has provided interesting results. The parametric analysis has highlighted in what percentage terms the steel increase, in terms of thickness and area of used strengthening profiles, allows an improvement of the RC column overall behaviour. It has also led towards a number of additional interesting considerations:

- The reinforcement with steel members gives a better resistance of the RC column. The performance increase is greater in terms of bending moment. In fact, in some models (i.e. those with batten plates height $h = 120$ mm and 160 mm) an increase of bending moment five times greater than that of the unreinforced column has been noticed.

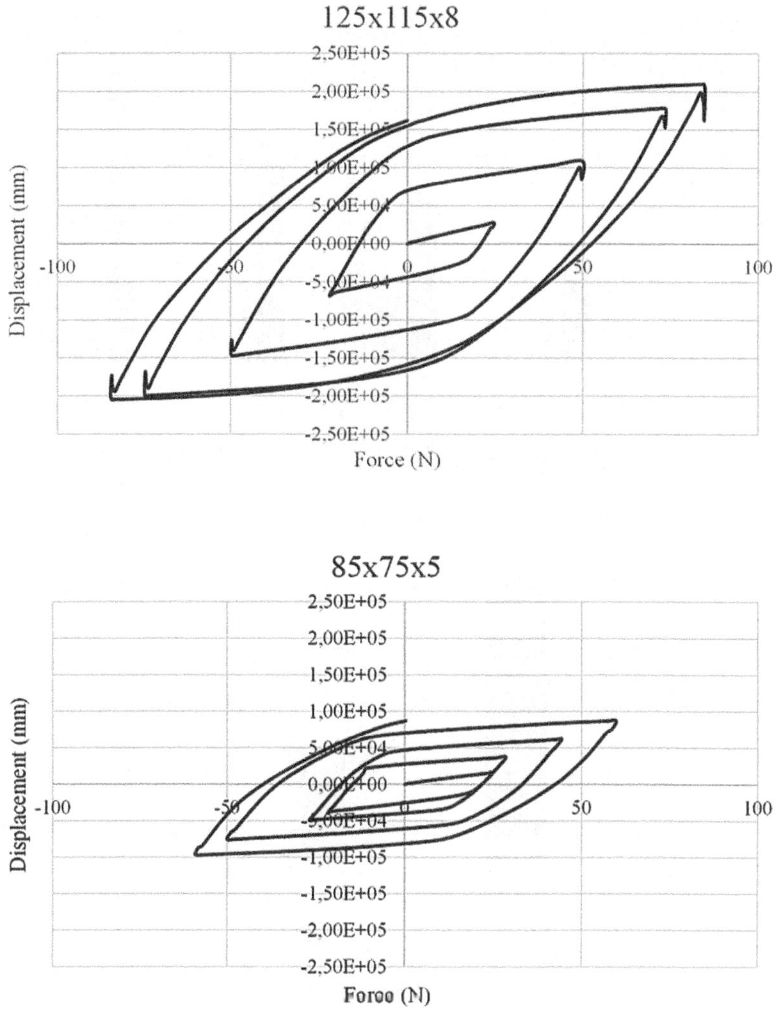

Fig. 8 Cyclic analysis results on the RC columns reinforced with the two optimal strengthening systems selected, namely L125 × 115 × 8 (**a**) and L85 × 75 × 5 (**b**)

- The major variation of the strengthened system resistance is derived from the changes of the three geometric parameters of steel members (height of batten plates *h*, length of angles *L* and thickness of batten plates and angles *t*).
- The normalisation of ultimate axial compression load and bending moment of the strengthened columns with respect to those of the original one and their representation as a function of different ratios among geometrical parameters of steel angles and batten plates (*h/t*, *h/L* and *L/t*) allow to achieve two optimal reinforcement solutions.

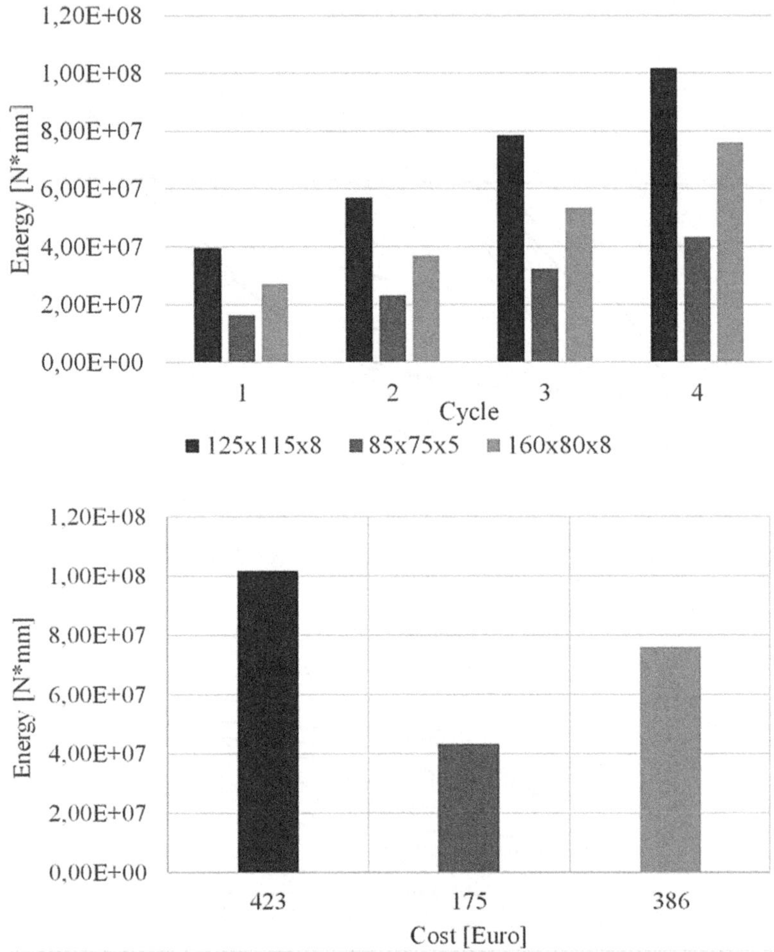

Fig. 9 Dissipated energy (**a**) and economic analysis (**b**) of three different types of steel reinforcing systems of RC columns

- The comparison between the two solutions in terms of both dissipated energy under cyclic horizontal actions and costs provides the intervention with 125 × 8 mm batten plates and 115 × 115 × 8 mm angles as the best strengthening system of the investigated RC column.

Acknowledgements The Author would like to acknowledge Eng. Antonio Davino for the precious contribution provided in writing Sects. 1, 2 and 3 of the paper. Also, the support of Eng. Domenico Longo, who performed FEM analyses within his MsC thesis, is gratefully acknowledged.

References

Aboutaha RS, Engelhardt MD, Jirsa JO et al (1996) Retrofit of concrete columns with inadequate lap splices by the use of rectangular steel jackets. Earthq Spectra 12:693–714
Aboutaha RS, Engelhardt MD, Jirsa JO et al (1999) Rehabilitation of shear critical concrete columns by use of rectangular steel jackets. ACI Struct J 96:68–78
Aboutaha RS, Engelhardt D, Jisra JO et al (1994) Seismic retrofit of R/C columns with inadequate lap splices. Proc ASCE '94 Structs Congr, Atlanta, Georgia
Adam JM, Ivorra S, Gimenéz E et al (2007) Behaviour of axially loaded RC columns strengthened by steel angles and strips. Steel Compos Struct 7:405–419
Akin A, Saka MP (2015) Harmony search algorithm based optimum detailed design of reinforced concrete plane frames subject to ACI 318–05 provisions. Comput Struct 147:79–95
Alcocer SM (1993) RC frame connections rehabilitated by jacketing. J Struct Eng 119:1413–1431
Alcocer SM, Jirsa JO (1993) Strength of reinforced concrete frame connections rehabilitated by jacketing. ACI Struct J 90:249–261
Braga F, Gigliotti M (2006) Analytical stress-strain relationship for concrete confined by steel stirrups and/or FRP jackets. J Struct Eng 32:1402–1416
Braga F, Gigliotti R, Laguardia R (2019) Intervention cost optimization of bracing systems with multiperformance criteria. Eng Struct 182:185–197
Braga F, Laterza M (1998) A new approach to the confinement of RC columns. Proc 11th European Conf on Earthq Eng, Rotterdam, The Netherlands
CEN—European Committee for Standardization (2005a) Design of composite steel and concrete structures. Part 1: General rules for buildings. Eurocode 4, Brussels, Belgium
CEN—European Committee for Standardization (2005b) Design of structures for earthquake resistance. Part 1: General rules, seismic actions and rules for buildings. Eurocode 4, Brussels, Belgium
Campione G, Cavaleri L, Di Trapani F, Ferrotto MF (2017) Frictional effects in structural behavior of no end-connected steel-jacketed RC columns: experimental results and new approaches to model numerical and analytical response. J Struct Eng. https://doi.org/10.1061/(ASCE)ST.1943-541X.0001796
Chaves LP, Cunha J (2014) Design of carbon fiber reinforcement of concrete slabs using topology optimization. Con Build Mater 73:688–98
Chisari C, Bedon C (2016) Multi-objective optimization of FRP jackets for improving the seismic response of reinforced concrete frames. Am J Eng Appl Sci 9:669–679
Dritsos SE (1997) Jacket retrofitting of reinforced concrete columns. Construction Repair 11:35–40
Dritsos S, Pilakoutas K (1994) Repair/strengthening techniques for structurally damaged RC columns. Proc 5th US National Conf on Earthq Eng, Chicago, Illinois
Dritsos S, Georgopoulos T, Pilakoutas K (1993) Experimental study on a composite technique for repair/strengthening techniques of RC columns. Proc 5th Intern Conf Struct Faults and Repair, Edinburgh, Scotland
Ezz-Eldeen HA (2016) Steel jacketing technique used in strengthening reinforced concrete rectangular columns under eccentricity for practical design applications. IJEET. https://doi.org/10.14445/22315381/IJETT-V35P243
Frangou M, Pilakoutas K, Dritsos S (1995) Structural repair/strengthening of RC columns. Constr Build Mater 9:259–266
Georgopoulos T, Dritsos SE (1994) A steel jacketing technique for strengthening RC columns. Proc 10th European Conf on Earthq Eng, Vienna, Austria
Ghimire A (2019) State-of-the-Art-review on masonry structure: identification behaviour and seismic retrofitting techniques for structures. IJSR 8:2319–7064
Ghobarah A, Tarek SA, Biddah A (1996) Seismic rehabilitation of reinforced concrete beam–column connections. Earthq Spectra 12:761–780
Greco R, Marano GC (2011) Optimal constrained design of steel structures by differential evolutionary algorithms. Int J Optim Civil Eng 3:449–474

Hibbitt, Karlsson & Sorensen, Inc., (HKS), ABAQUS Standard/explicit user's manual, Version 6.7, USA, 2008
Lavan O, Dargush GF (2009) Multi-objective evolutionary seismic design with passive energy dissipation systems. J. Earthq. Eng. https://doi.org/10.1080/13632460802598545
Li YF, Chen SH, Cheng KC et al (2005) A constitutive model of concrete confined by steel reinforcements and steel jackets. Can J Civ Eng 32:279–288
Lippi FV, Mangini F, Salvatore W et al (2018) Steel in Structural Interventions on Existing Masonry and Reinforced Concrete Buildings: Examples and Applications (in Italian). Dario Flacco Editor, Palermo, Italy
Liu M, Burns SA, Wen YK (2003) Optimal seismic design of steel frame buildings based on life cycle cost considerations. Earthquake Eng Struct Dyn 32:1313–1332
Lyn A, Moehle JP, Mahin SA et al (1996) Seismic evaluation of existing reinforced concrete building columns. Earthq Spectra 12:715–739
Masri AC, Goel SC (1996) Seismic design and testing of an RC slab–column frame strengthened by steel bracing. Earthq Spectra 12:645–666
Ministry of Infrastructure and Transport (2008) Updating technical standards for construction (in Italian), Official Gazette, Rome (nr. 29 of 04-02-2008)
Ministry of Infrastructure and Transport (2018) Technical standards for construction (in Italian), Official Gazette, Rome (nr. 42 of 20-2-2018)
Mitropoulou CC, Lagaros ND, Papadrakakis M (2011) Life-cycle cost assessment of optimally designed reinforced concrete buildings under seismic actions. Reliab Eng Syst Saf 96:1311–1331
Montuori R, Piluso V (2009) Reinforced concrete columns strengthened with angles and battens subjected to eccentric load. Eng Struct 31:539–450
Murty CVR, Greene M, Jain SK et al (2005) Earthquake Rebuilding in Gujarat, India—a recovery reconnaissance report. Earthquake Engineering Research Institute, Oakland, CA
Nagaprasad P, Sahoo DR, Rai DC (2009) Seismic strengthening of R.C. columns using external steel cage. Earthquake Eng Struct Dyn 38:1563–1586
Nunziata V (2011) Theory and Practice of Steel Structures (in Italian). Dario Flaccovio Editore, Palermo
Papavasileiou GS, Charmpis DC (2016) Seismic design optimization of multi–storey steel–concrete composite buildings. Comp Struct 170:49–61
Paulay T, Priestley MJN (1992) Seismic design of reinforced concrete and masonry buildings. Wiley, New York
Priestley MJN, Seible F, Xiao Y et al (1994) Steel jacket retrofitting of reinforced concrete bridge columns for enhanced shear strength—part 1: theoretical considerations and test design. ACI Struct J 91:394–405
Ramirez JL (1996) Ten concrete column repair methods. Constr Build Mater 10:195–202
Raza S, Khan MKI, Menegon SJ et al (2019) Strengthening and repair of reinforced concrete columns by jacketing: state-of-the-art review. Sustainability. https://doi.org/10.3390/su11113208
Razvi SR, Saatcioglu M (1992) Strength and ductility of confined concrete. J Struct Eng 125:281–298
Rodriguez M, Park R (1991) Repair and strengthening of reinforced concrete buildings for seismic resistance. Earthq Spectra 7:439–459
Sakino K, Sun Y (2000) Steel Jacketing for Improvement of Column Strength and Ductility. Proc 12th World Conf on Earthq Eng, Auckland, New Zealand
Seible F, Priestley MJN, Hegemier GA et al (1997) Seismic retrofit of RC columns with continuous carbon fiber jackets. J Compos Constr 1:52–62
Tamai S, Sato T, Okamoto M (2000) Hysteresis model of steel jacketed RC columns for railway viaducts. Proc 16th Congress of IABSE, Lucerne, Switzerland
Di Trapani F, Malavisi M, Marano GC et al (2020) Optimal seismic retrofitting of reinforced concrete buildings by steel-jacketing using a genetic algorithm-based framework. Eng Struct. https://doi.org/10.1016/j.engstruct.2020.110864

Unjoh S, Kawashima K (1992) Seismic inspection and seismic strengthening of reinforced concrete bridge piers. Proc 10th World Conf on Earthq Eng, Madrid, Spain

Valluvan R, Kreger ME, Jirsa JO (1993) Strengthening of column splices for seismic retrofit on non-ductile reinforced concrete frames. ACI Struct J 90:432–433

Wasti ST, Ozcebe G (2006) Advances in earthquake engineering for urban risk reduction. Nato Science Series: IV: Earth and Environmental Sciences, Springer, Netherlands

Wu YF, Liu T, Oehlers DJ (2006) Fundamental principles that govern retrofitting of reinforced concrete columns by steel and rp jacketing. Adv Struct Eng. https://doi.org/10.1260/136943306778812769

Zhong Y, Ren F, Tian J (1986) Repair and strengthening of reinforced concrete columns. Proc US PRC Joint Workshop on Seismic Resistance of Masonry Struct, Harbin, China

Zou XK, Chan CM, Li G et al (2007) Multiobjective optimization for performance based design of reinforced concrete frames. J Struct Eng 133:1462–1474

Chloride Profile Modeling for Reinforced Concrete Structures Present in Marine Environment

C. E. T. Balestra, T. A. Reichert, A. L. P. Vizzotto, and G. Savaris

Abstract The durability and service life of reinforced concrete structures in the marine environment have been the subject of several studies in the literature, due to the great importance for port infrastructure and the high level of aggressiveness offered by this environment to reinforced concrete structures, mainly by the chloride ions action in the corrosion of the reinforcement. In this way, the present chapter deals with the modeling of chlorides concentration profiles, where an analysis recently developed for application of the Modified Holliday Equation is presented. For this, different profiles models were performed, representing different materials and environmental conditions using this new methodology. The results showed that the model based on the Modified Holliday Equation can represent the diffusion zone for structures with different characteristics (materials, environmental, age and location), and is also able to represent chloride concentrations in the convection zone, allowing to estimate the service life of structures built in marine environments around the world.

Keywords Chloride profile · Reinforced concrete structures · Reinforcement corrosion · Modelling · Service life · Durability

1 Introduction

The durability of reinforced concrete infrastructure in the marine environment has been the subject of several researches around the world, considering that (1) the marine infrastructure modal plays a fundamental role in trade relations between

C. E. T. Balestra (✉) · T. A. Reichert · A. L. P. Vizzotto · G. Savaris
Federal University of Technology - Paraná, Curitiba, Brazil
e-mail: carlosbalestra@utfpr.edu.br

T. A. Reichert
e-mail: thiago.reichert@outlook.com

A. L. P. Vizzotto
e-mail: andre-vizzotto@hotmail.com

G. Savaris
e-mail: gsavaris@utfpr.edu.br

J. M. P. Q. Delgado (ed.), *Durability of Concrete Structures*, Building Pathology and Rehabilitation 16, https://doi.org/10.1007/978-3-030-62825-3_3

different nations, acting in 90% of operations of export, besides that several capitals and important cities are located in coastal regions and their structures are subject to degradation by the marine environment; (2) the degradation of reinforced concrete structures present in a marine environment involves high rehabilitation costs, directly in the repair of the structure and also indirectly in the idle time of commercial activity; (3) the corrosion of the reinforcements resulting from the action of chlorides is the main pathological manifestation noted, which can occur at high speeds and compromise partially or fully the use and safety of these structures (Mehta and Monteiro 2006; Balestra et al. 2019d; Apostolopoulos et al. 2013; Neville and Brooks 2010; Balestra et al. 2016).

To understand the corrosion process of reinforcement in concrete, firstly is necessary to point out that the alkaline pH of the solution present in the concrete pores promotes an adequate environment for the formation of a passive film, that covers and protects the reinforcement inside the concrete. However, in the case of the marine environment, chlorides penetrate through the concrete pore network and when reach the reinforcement region with concentration greater than 0.4% in relation to the cement mass, create conditions for depassivate the reinforcement and initiate an autocatalytic corrosion process The corrosion caused by the chlorides is known as pitting, and they are characterized by corrosion points that deepen in the reinforcement cross section along the corrosion process. Figure 1 shows a pit on the metallic surface of the reinforcement. (Mehta and Monteiro 2006; Neville and Brooks 2010; Balestra et al. 2016; François et al. 2013; Apostolopoulos 2009).

The pitting corrosion process is presented in Fig. 2 (Mehta and Monteiro 2006; Neville and Brooks 2010; Broomfield 2007). When chlorides reach the reinforcement occur the acidification of the medium, that destroys the passive film in this region. After the depassivation, the chlorides end up removing the metallic iron from the reinforcement, through the formation of $FeCl_2$, which is dissociate in the medium, allowing (1) the combination of the metallic iron with hydroxyl ions (OH-) present in the medium to the formation of iron hydroxides ($Fe(OH)_2$), known as corrosion products; (2) the return of chlorides to the pits for new removals of metallic iron in

Fig. 1 Steel reinforcement with pitting corrosion (Balestra et al. 2018)

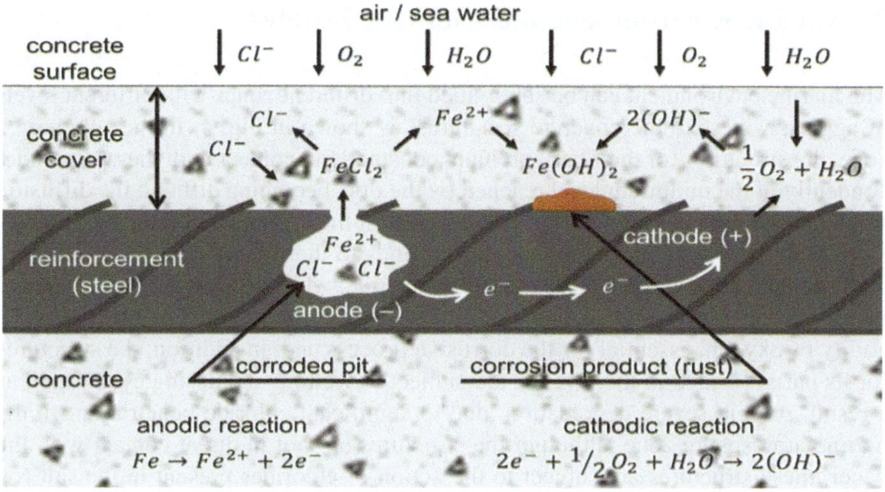

Fig. 2 Reinforced concrete corrosion scheme due to chlorides action (Mehta and Monteiro 2006; Neville and Brooks 2010; Broomfield 2007)

a self-catalytic process. From this corrosion process, a significant reduction of the active area of steel reinforcement can be observed, increasing the effective stress, and affecting the mechanical properties of the reinforcement, mainly related to ductility, as showed in Fig. 3 (Balestra et al. 2018).

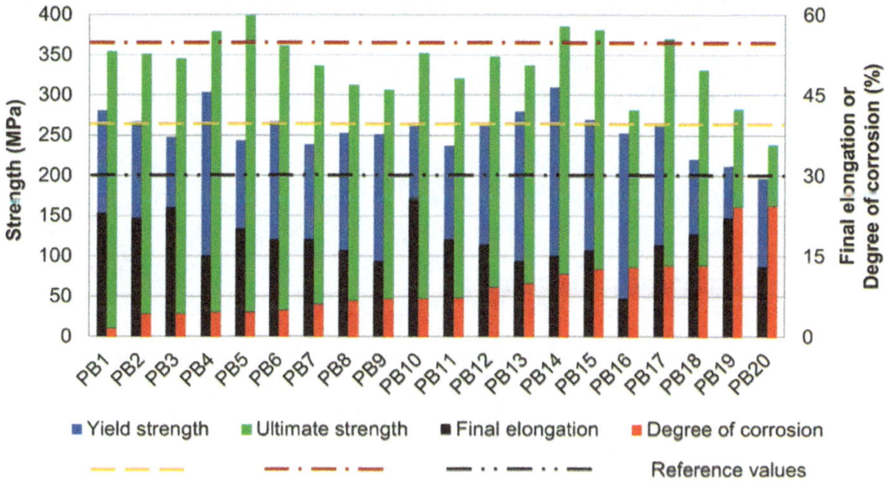

Fig. 3 Relationship between loss of mass and yield strength, ultimate strength and final elongation of corroded reinforcement (Balestra et al. 2018)

2 Marine Environment and Chloride Profiles

The marine environment can be subdivided into different zones with different levels of aggressiveness to the concrete structures, as shown in Fig. 4. In the submerged zone, the structures (or their parts) remain constantly in contact with the water, independently of the minimum level reached by the tide, becoming difficult the diffusion of oxygen to the environment for participation in the corrosion process of the reinforcement. The tidal variation zone corresponds to the region located between the maximum and minimum levels reached by the tide, while the splash zone is located just above the upper limit reached by the tide. In both zones there is significant availability of oxygen, essential for the corrosive process and, in addition, the structures (or its parts) located in these zones are subject to wetting and drying cycles and, as a result, present severe degradation, due to significant chloride penetration. In the marine atmosphere zone, although the structures are not in direct contact with the water, these structures are subject to the action of chlorides present in the salt fog formed from the breaking waves. This salt fog can be carried and reach structures that are located up to 1 km from the seacoast (Mehta and Monteiro 2006; Neville and Brooks 2010; Balestra et al. 2019a, b, 2020).

The entry of chlorides into the concrete can be evaluated using chloride profiles (Fig. 5), that represent the variation in chloride concentration at different depths from the concrete surface. As shown in Fig. 5, in the chloride concentration profiles is possible to observe the existence of a peak of maximum chloride concentration, delimiting a region between the surface and the peak, known as the convection zone, and an area after the peak, known as the diffusion zone.

The convection zone is submitted to wetting and drying cycles, with capillary absorption being the predominant mechanism for the penetration of chlorides. In the case of the diffusion zone, the diffusion mechanism is governed predominantly by the Fick's Laws of diffusion. The depth of the peak chloride concentration depends

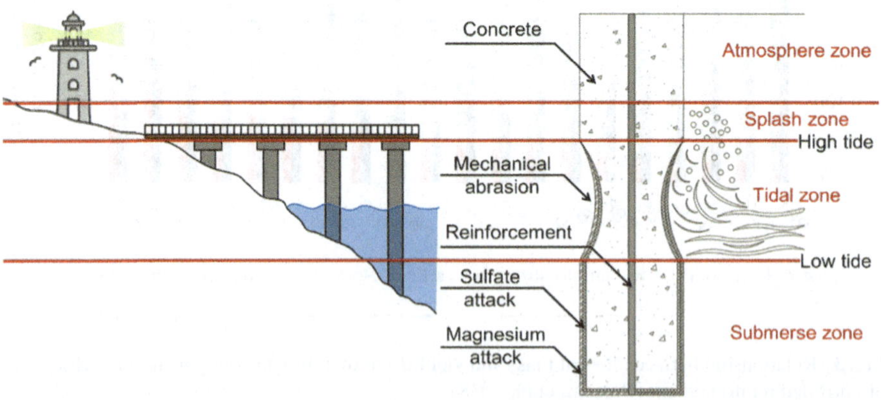

Fig. 4 Representation of marine aggressiveness zones. Adapted from (Mehta and Monteiro 2006)

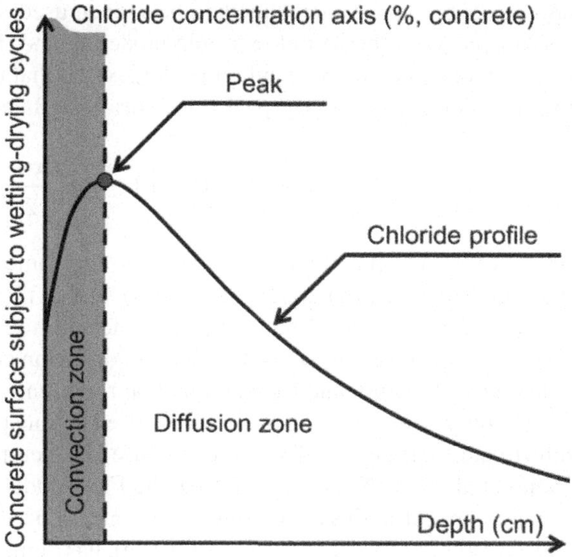

Fig. 5 Chloride profile representation (Balestra et al. 2019a)

mainly on the pore microstructure of the concrete, with depths between 0.5 to 2 cm, observed in usual structures. The maximum concentration of chlorides at the peak is also dependent on the characteristics of the concrete and its porosity, being observed maximum values of 1.5% in relation to the cement mass, for usual concrete structures aged over 30 years with cement consumption up to 400 kg/m^3 (Castro et al. 2001).

In order to determinate a profile, chloride concentrations can be quantified by extracting samples of powdered concrete, obtained at different depths, analyzed using potentiometric titration, as recommended by ASTM C 1152 (2012), however, recent studies used the X-Ray Fluorescence Spectroscopy technique to preserve the powdered concrete sample (Balestra et al. 2019a, b, c, 2020).

3 Methods of Modeling Chloride Concentration Profiles for Durability Studies

Considering a defined chloride profile, it is possible to estimate service life for concrete structures. In this case, the most common method of modeling is to define the position of the chloride concentration peak and proceed to modeling only the diffusion zone, using Fick's laws of diffusion. In this case, it is considered, as an initial condition, that the initial concentration of chlorides in the concrete is zero at the initial time and, as a boundary condition, that the concentration of chlorides on the surface Cs remains constant over time. In this way, the solution to the Fick's 2nd Law (Eq. 1) has been used in models for predicting the service life of concrete structures in different areas of marine aggressiveness. In this context, regression analyzes and use of the least squares method have been used, in order to adjust the diffusion coefficient (D) to perform the profile modeling and, for this, the concentration axis has been relocated to the depth of peak (Δx), so that the surface concentration (Cs), in Eq. 1, corresponds to the peak concentration (Balestra et al. 2019a).

$$C(x,t) = Cs \cdot \left[1 - \operatorname{erf}\left(\frac{x - \Delta x}{2\sqrt{Dt}}\right)\right] \tag{1}$$

where $C(x, t)$ is the chloride concentration, at time t, at a deep x; x is the depth measured from the concrete surface (cm); Δx is the peak depth, in relation to the concrete surface (cm); Cs is the peak concentration (considered as the superficial chloride concentration); D is the apparent diffusion coefficient (cm^2/s); erf(z) is the Gauss error function and t is the time that represents the structure age (s).

Although this modeling procedure is used by most studies presented in the literature, related to the durability and service life of concrete structures (Pang and Li 2016; Vieira et al. 2018; Wang et al. 2018), the Fick model is unable to adequately represent the concentrations in the convection zone. Thus, the Modified Holliday Equation (Eq. 2), proposed by Balestra et al. (2019a), has being used recently, to represent the concentration of chlorides at both, the convection zone and the diffusion zone. In this case, considering the parameters of the chloride profile and structure: (1) Chloride concentrations at different depths, (2) peak position and (3) peak chloride concentration and, finally, (4) age of the structure, in years) it is possible to use regression techniques to determine the diffusion coefficient (D), and perform the modeling to estimate the variation of chloride concentration in the concrete for different ages. In this case, it should be noted that the application of the Modified Holliday Equation requires the presence of a peak of maximum concentration of chlorides, which are generally observed in structures older than 20 years.

$$C(x,t) = \frac{1/R_1}{\left[1 + \frac{(x-R_3)^2}{D \cdot t}\right]} \tag{2}$$

where $1/R_1$ corresponds to the concentration peak; x is the depth measured from the concrete surface (cm); D is the apparent diffusion coefficient (cm^2/s); R_3 corresponds to the peak depth, in relation to the concrete surface (cm); t is the structure age (s).

Figure 6 presents a comparison between models based on the solution of the Fick's 2nd Law of diffusion and based on the Modified Holliday Equation. Balestra et al. (2019a) shown that the solution of Fick's second law is able to model only the diffusion zone data and that, when extrapolating the model to the convection zone, the model clearly does not represent the concentration data. On the other hand, using the Modified Holliday Equation it was possible to model the chloride concentration data for both zones. Must be highlighted that these data come from a real structure that remained under natural degradation in a tidal zone for more than 50 years. At this point, even extrapolating the model to an age of 100 years, it is possible to observe

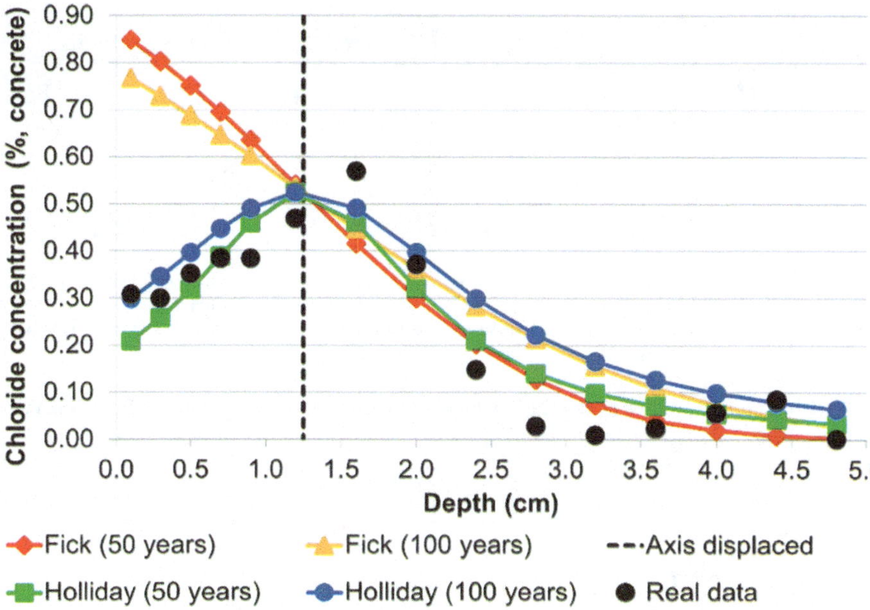

Fig. 6 Chloride profiles modeled using Fick's 2nd law and Modified Holliday Equation (Balestra et al. 2019a)

that both the Fick model (in the diffusion zone) and the Modified Holliday model (in both zones) are capable of estimate the chloride concentration in the diffusion zone for more advanced ages.

In the design of the Modified Holliday Equation, the model was established so that the denominator of Eq. 2 is dimensionless. This means that, regardless of the units used for the chloride concentration data, since the denominator of Eq. 2 remains dimensionless, the model can be applied. In fact, it is observed that the Holliday model can be used for modeling regardless of whether the concentration is expressed in terms of the chloride concentration in relation to the concrete mass or in relation to the cement mass, independent of the peak position (superficial or deep), of the area of marine aggressiveness where the structure is found, the characteristics of the concrete (with respect to the water/cement ratio, cement and aggregates consumption) and the global location and age of the structure. Figures 7, 8 and 9 present chloride profiles obtained in different structures around the world and modeled after the solution of the Fick's 2nd Law and the Modified Holliday Equation, where the potential for using the Holliday methodology, proposed by Balestra et al. (2019a), is clearly observed.

The profiles presented in Figs. 7, 8 and 9 were obtained from structures present in different locations around the world and different zones of marine aggressiveness, with ages varying between 10 and 50 years. In this sense, Fig. 7 presents profiles obtained from structures located on the south coast of China, aged between 18 and 21 years old, present in a splash zone (Tang 2003); Fig. 8 presents profiles of structures

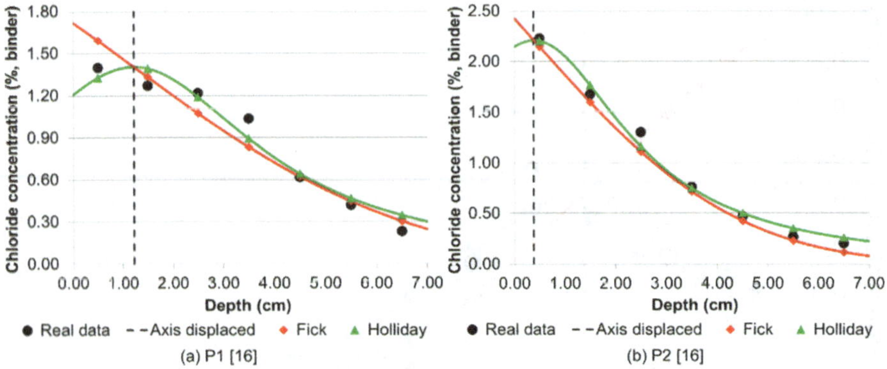

Fig. 7 **a**, **b**. Chloride profiles modeled using Fick's 2nd law and Modified Holliday Equation. Pang and Li (2016)

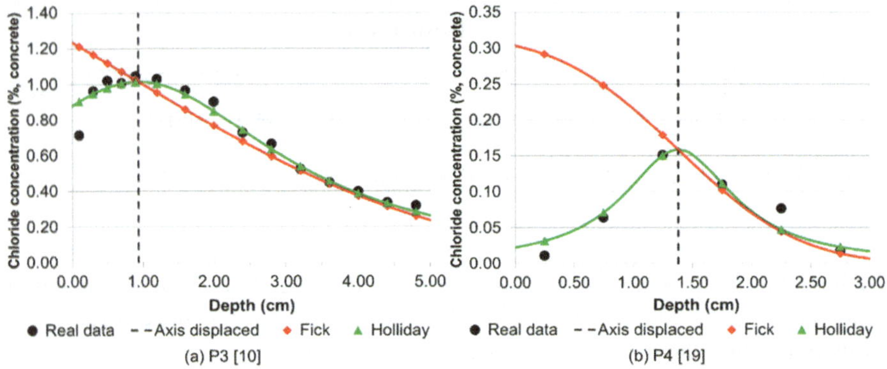

Fig. 8 **a** Chloride profile modeled using Fick's 2nd law and Modified Holliday Equation (Balestra et al. 2019a) **b** Chloride profile modeled using Fick's 2nd law and Modified Holliday Equation (Nunes and Guimarães 2006)

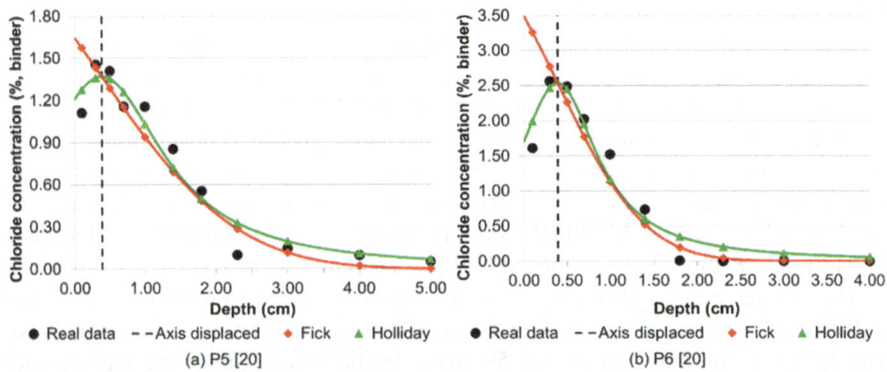

Fig. 9 **a**, **b** Chloride profiles modeled using Fick's 2nd law and Modified Holliday Equation (Tang 2003)

located on the coast of Brazil, but in different places, where the first (Fig. 8a), aged 50, is located in a tidal range (Balestra et al. 2019a), while the second (Fig. 8b), aged 10, is located in a marine atmosphere zone (Nunes and Guimarães 2006); and, finally, Fig. 9 shows profiles obtained from Swedish structures, with 10 years of exposure and present in a marine atmosphere zone (Fig. 9a) and submerged zone (Fig. 9b) (Tang 2003). In addition to differences in the area of marine aggressiveness, location and age of structures, chloride concentration profiles were obtained from structures that have different concrete characteristics (cement consumption and water/cement factors) as shown in Table 1.

In a general analysis of the chloride concentration profiles, highlights the behavior of the models in the convection zone (located between the surface and the peak of concentration). In this region, capillarity absorption is the predominant mechanism of aggressive ions ingress into the concrete, and it is possible to observe that the Modified Holliday model is able to model the data present in the convection zone, contrary to what is observed when applied Fick's Law. In this regard, Fick's laws deal exclusively with the mechanism of ionic diffusion, thus, it is possible to justify the reason why the model based on the solution of Fick's second law fits well only the data after the peak, located in the diffusion zone. In addition, in the diffusion zone, it is possible to observe that both models have a great capacity to represent the chloride concentration profiles, even when extrapolations of time are performed, as shown in Fig. 6.

One of the main difficulties in adopting the Fick's model to analyze the penetration of chlorides in concrete is related to the necessity of displacement of the axis of chloride concentrations to the peak position, and so that model contemplates only the data in the diffusion zone, resulting, after the adjustment, an erroneous extrapolation of the results for the convection zone. However, discarding the points located before the peak of concentration leads to errors in the values of the model parameters, in addition to the results not representing the concentration of chlorides in the convection zone in an acceptable manner. In this sense, the model from the Modified Holliday Equation has a great advantage in allowing the analysis of the convection zone, since the Holliday model can fully describe the variation in the chlorides concentration in the concrete throughout the entire profile over time. The importance of using the points located in the convection zone occurs since this zone is responsible for the supply of chlorides to the diffusion zone, and becomes more important for profiles in which there is a low number of points after the peak, i.e. the P4 profile (Fig. 8b), in which the Fick's model, discarding the points in the convection zone, becomes practically inapplicable.

On the other hand, the Holliday model, as well as the Fick's model, depends fundamentally on the determination of the peak position and peak concentration of chlorides in the profile, since the parameters of the models are based on these measurements. However, it has been observed, according to the study by Balestra et al. (2019a), that the Modified Holliday Equation can determine such parameters, through the variables R_1 and R_3 of the model, using only the least squares method for its adjustments, even if the peak is located a few millimeters depth from the

Table 1 Characteristics of concrete structures analyzed at Figs. 7, 8 and 9

Profile	References	Marine agr. zone				Location	Age (years)	Concrete characteristics	
		SubZ	TD	SZ	AtmZ			Cement consumption (kg/m^3)	w/c ratio
P1	Pang and Li (2016)			X		China	18	–	0.45
P2	Pang and Li (2016)			X		China	21	–	0.45
P3	Balestra et al. (2019a)		X			Brazil	50	440	–
P4	Nunes and Guimarães (2006)				X	Brazil	10	–	–
P5	Tang (2003)				X	Sweden	10	420	0.40
P6	Tang (2003)	X				Sweden	10	500	0.30

Legend: SubZ = Submerged zone; TD = Tidal Zone; SZ = Splash Zone; AtmZ = Atmospheric zone concrete surface; while the Fick's model requires, in most cases, that the researcher intervene and empirically define a peak position to adjust the characteristics of the concrete.

4 Conclusions

This chapter aims to present methods of modeling chloride concentration profiles, focusing to the durability and service life of concrete structures present in the marine environment. The main conclusions are:

- The modeling of chloride concentration profiles allows the analysis of the service life of concrete structures present in different areas of marine aggressiveness, providing important information for managers to make decisions about when to carry out maintenance or rehabilitation of structures.
- The corrosion process resulting from the action of chloride ions present in the marine environment leads to a decrease in the resistance of the steel bars, since pitting corrosion causes localized damages that reduces the reinforcement cross section.
- The modeling of chloride profiles is commonly performed through the Fick's 2nd Law of diffusion; however, such methodology does not apply to the representation of chloride concentrations in the convection zone, which generates modeling difficulties and predictions of the service life of concrete structures. Thus, one way to represent the chloride concentrations present simultaneously in the convection and diffusion zones is through the Modified Holliday Equation.
- The modified Holliday equation can be applied to any chloride profile, as long as there is a peak (even if close to the surface), regardless of age, location (geographic and/or in the areas of marine aggressiveness) and concrete characteristics (cement consumption and water/cement ratio).

References

American Society for Testing and Materials (2012) ASTM C1152: 2012: standard test method for acid-soluble chloride in mortar and concrete. West Conshohocken

Apostolopoulos CA (2009) The influence of corrosion and cross-section diameter on the mechanical properties of B500c steel. J f Mater Eng Perform 18(2):190–195

Apostolopoulos CA, Demis S, Papadakis VG (2013) Chloride-induced corrosion of steel reinforcement: mechanical performance and pit depth analysis. Constr Build Mater 38:139–146

Balestra CET et al (2016) Corrosion degree effect on nominal and effective strengths of reinforcement naturally corroded. J Mater Civil Eng 28(10):04016103

Balestra CET, Lima MG, Mendes AZ, Medeiros-Junior RA (2018) Effect of corrosion degree on mechanical properties of reinforcements buried for 60 years. IBRACON Struct Mater J 11(3):474–498

Balestra CET, Nakano AY, Savaris G, Medeiros-Junior RA (2019a) Reinforcement corrosion risk of marine concrete structures evaluated through electrical resistivity: proposal of parameters based on field structures. Ocean Eng 187:106167

Balestra CET, Reichert TA, Pansera WA, Savaris G (2019b) Chloride profile modeling contemplating the convection zone based on concrete structures present for more than 40 years in different marine aggressive zones. Constr Building Mater 198:345–358

Balestra CET, Reichert TA, Savaris G (2019c) Contribution for durability studies based on chloride profiles analysis of real marine structures in different marine aggressive zones. Constr Building Mater 206:140–150

Balestra CET, Savaris G, Nascimento RS, Medeiros Jr RA (2019d) Relations between chloride determination in real concrete structures by X-ray Fluorescence spectroscopy and by potentiometric titration. J Build Pathol Rehab 4(18)

Balestra CET, Reichert TA, Pansera WA, Savaris G (2020) Evaluation of chloride ion penetration through concrete surface electrical resistivity of field naturally degraded structures present in marine environment. Constr Building Mater 230:1169789

Broomfield JP (2007) Corrosion of steel in concrete: understanding, investigation and repair. Taylor & Francis, London, p 294

Castro P, De Rincon OT, Pazini EJ (2001) Interpretation of chloride profile from concrete exposed to tropical marine environments. Cement Concrete Res 31:529–537

François R, Khan I, Dang VH (2013) Impact of corrosion on mechanical properties of steel embedded in 27-year-old corroded reinforced concrete beams. Materi Struct 46(6):889–910

Mehta PK, Monteiro PJM (2006) Concrete: microstructure, properties and materials. 3th edition. Mc-Graw Hill. New York, 674 p

Neville AM, Brooks JJ (2010) Concrete technology. 2st Ed. Pearson. London. 448 p

Nunes JLO, Guimarães ATC (2006) Seawater distance- Factor to be considered in the chloride attack intensity. II Ocean Engineering Workshop Federal University of Rio Grande, 2006. Available in: http://repositorio.furg.br/handle/1/1644?fbclid=IwAR3s6cYJ2pkdy-SgwVwiEN_13b9iVSy5hFnmcqvH7CqKC33i1zRVgtWsC1k (In Portuguese)

Pang L, Li Q (2016) Service Life prediction of RC structures in marine environment using long term chloride ingress data: comparison between exposure trials and real structures surveys. Constr Building Mater 113:979–987

Tang L (2003) Chloride ingress in concrete exposed to marine environment—Field data up to 10 years exposure. Swedish National Testing and Research Institute. Building Technology and Mechanics. SP REPORT 2003:16. 68 p. Available in: http://www.diva-portal.org/smash/get/diva2:962253/FULLTEXT01.pdf?fbclid=IwAR3Jx1g8JNEkpCexF6El0q8bM380MfDVDZDGy-j_rSf2EoFCDYdTtexPgRM

Vieira DR, Moreira ALR, Calmon WK, Dominicini WK (2018) Service life modeling of a bridge in a tropical marine environment for durable design. Constr Building Mater 163:315–325

Wang Y, Wu L, Wang Y, Li Q, Xiao Z (2018) Prediction model of long-term chloride diffusion into plain concrete considering the effect of heterogeneity of materials exposed to marine tidal zone. Constr Building Mate 1593:297–315

Concrete Electrical Resistivity as a Durability and Service Life Parameter for Reinforced Concrete Structures

C. E. T. Balestra, T. A. Reichert, and G. Savaris

Abstract Electrical resistivity becomes one of the main non-destructive techniques in the evaluation of the service life and durability of reinforced concrete structures, considering chloride penetrability and/or the probability of reinforcement corrosion, being pointed out by different standard and studies in literature. Due to the use of a non-destructive methodology of easy operation, reliable, with expeditious results and low cost, it becomes an important tool for decision making by managers responsible for plans and strategies for concrete structures maintenance. Thus, this chapter presents definitions and aspects of electrical resistivity in concrete, test procedures and interpretation of results and, finally, the parameters that most influence concrete resistivity are discussed.

Keywords Electrical resistivity · Reinforced concrete structures · Reinforcement corrosion · Nondestructive tests · Service life · Durability

1 Introduction

The development of sustainable and durable concrete structures puts forward studies in the field of Civil Engineering, considering that (1) the construction industry is responsible for the consumption of approximately 75% of the planet's natural resources and (2) the high maintenance and rehabilitation costs of degraded structures requires planning with the managers responsible for these works. Besides maintenance and rehabilitation costs, frequently the access to concrete structures (or their parts) for the diagnosis of pathologies is an arduous task, i.e. structural elements of bridges or parts of maritime infrastructure present in several different areas of marine

C. E. T. Balestra (✉) · T. A. Reichert · G. Savaris
Federal University of Technology - Paraná, Curitiba, Brazil
e-mail: carlosbalestra@utfpr.edu.br

T. A. Reichert
e-mail: thiago.reichert@outlook.com

G. Savaris
e-mail: gsavaris@utfpr.edu.br

J. M. P. Q. Delgado (ed.), *Durability of Concrete Structures*, Building Pathology and Rehabilitation 16, https://doi.org/10.1007/978-3-030-62825-3_4

aggressiveness, which can involve since a simple visual inspection to the extraction of specimens for analysis in the laboratory (Mehta and Monteiro 2006; Balestra et al. 2019; Akhtar and Sarmah 2018; Neville and Brooks 2010).

As described in medicine, to a prophylaxis or intervention be proposed to treat a set of pathological problems, a diagnosis through exams about the concrete structure is necessary. In this context, the adoption of less invasive techniques to structures, known as non-destructive tests (NDT) has been one of the main lines of research about durability of concrete structures, since they allow to obtain results quickly, reliable and easy to evaluate and correlate. For this reason, one of the main techniques used recently deals with the analysis of electrical (ionic) resistivity of concrete, correlating this property with the probable chloride penetration rate and/or with the corrosion probability (Mehta and Monteiro 2006; Balestra et al. 2019; Akhtar and Sarmah 2018; Neville and Brooks 2010; Nguyen et al. 2017, 2018; Chen et al. 2014; Sengul 2014).

Electrical resistivity (or, in this case of concrete, ionic resistivity) is an intrinsic characteristic of the material, related to its ability to oppose the passage of electrical current, being governed by Ohm's law, presented in Eq. 1. When the material is subjected to a potential difference, an electric current characterized by the free flow of electrical charges is established and, in this case, free electrons collide with each other and against atoms of the conductor, making the passage of electric current difficult, which is called electrical resistance. Thus, the nature of the material influences on resistance, so that, as greater is the amount of free electrons, easier will be to establish an electric current. This specific characteristic is known as electrical resistivity (Nguyen et al. 2017, 2018; Chen et al. 2014; Sengul 2014).

$$R = \frac{V}{I} \tag{1}$$

where R is the electrical resistance of the concrete (Ω); I the electric current flowing through the circuit (A) and V is the potential difference between the electrodes (V).

Once determined the electrical resistance of the material, the electrical resistivity can be calculated using Eq. 2, where ρ is the electrical resistivity, in Ω.m; k_f is a form fator, associated to the sample size, in m; and R is the electrical resistance, in Ω.

$$\rho = \mathrm{k_f} \cdot \mathrm{R} \tag{2}$$

Concrete can present different values of electrical resistivity, depending on the degree of pores saturation. Thus, the moisture present in the pores can act as a semiconductor, with electrical resistivity of the order of up to 10^2 Ω.m, whereas, when dried in a kiln, the electrical resistivity can be of the order of 10^9 Ω.m (Neville and Brooks 2010).

2 Electrical Resistivity Determination Methods

Two are the most used methodologies for determining the electrical resistivity of concrete: First, the direct method, presented in the standards ASTM C1760 (ASTM 2012), AASHTO T358-15 (AASHTO 2015) and UNE 83988-1 (AENOR 2008), and the second, the Wenner method (four-point method), presented in the standards AASHTO T358-15 (AASHTO 2015) and UNE 83988-2 (AENOR 2014). In this case, it is worth mentioning that, although electrical resistivity is a property of the material, factors such as frequency and nominal voltage may vary between standards.

The direct method allows the determination of the electrical resistivity by measuring the potential difference and the electric current flowing through a sample under the action of an electric field applied as shown in Fig. 1. For application of this test method, the concrete specimens must be cylindrical, with a diameter of 100 mm and length between 100 and 200 mm, with the form factor (K_f) calculated by the ratio of the cross-sectional area of the specimen (S, in m^2) and its length (L, in m). Thus, the electrical resistivity, in the direct method, can be calculated using Eq. 3. As a disadvantage, the use of direct current in this method can lead to measurement errors due to the polarization of the electrodes.

$$\rho = R \cdot \frac{S}{L} = \frac{V}{I} \cdot \frac{\pi d^2}{4L} \tag{3}$$

The Wenner method consists of measuring the electrical resistivity through the contact of four probes with the concrete specimen, where the external probes apply an electric current (I) and the internal probes monitor the potential difference (V), as shown in Fig. 2. In this method, for analysis of specimens with semi-infinite dimensions, the form factor (K_f) depends on the spacing between the probes (a, in m), usually adopted values of 38 or 50 mm, and thus, the electrical resistivity by the Wenner method, can be calculated using Eq. 4.

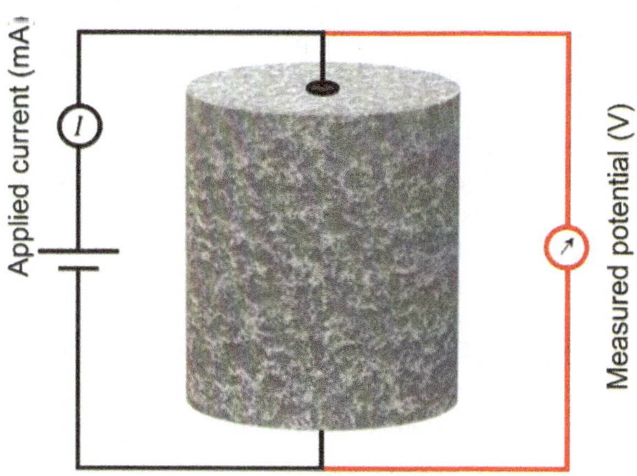

Fig. 1 Direct method scheme to electrical resistivity determination of concrete specimens

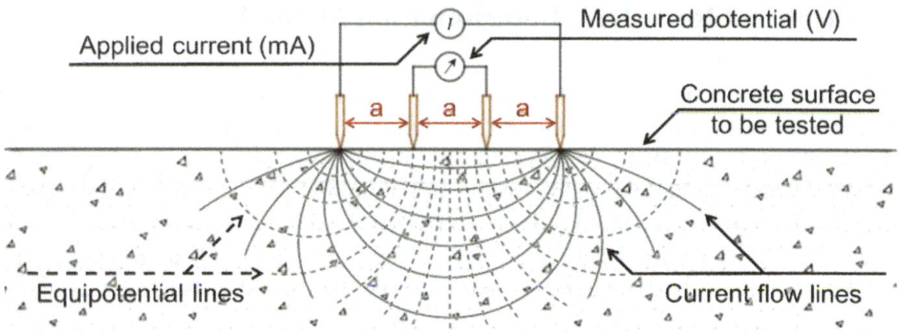

Fig. 2 Wenner method scheme for concrete electrical resistivity determination

$$\rho = \frac{V}{I} \cdot 2\pi a \tag{4}$$

The Wenner method has advantages over the direct method, since the results are expeditious and the equipment is portable, allowing the use in field structures, avoiding samples extraction. However, two observations should be emphasized: (1) This method performs surface resistivity measurements, which may be different from the resistivity of the interior of the structure, since the degree of pore saturation on the surface and inside the concrete can be different; (2) The presence of reinforcement can affect the results of resistivity in the field, since the bars act as conductors, therefore, it is recommended to use a pacometer for field measurements, in order to detect the location of the reinforcement and avoid incorrect positioning of the probes.

The results obtained from the electrical resistivity of the concrete have been used for analyzes of durability and service life of reinforced concrete structures, relating this parameter to the corrosion of the reinforcements, since the corrosion phenomenon has an electrochemical character (Mehta and Monteiro 2006; Balestra et al. 2019; Neville and Brooks 2010). In this way, two correlation methods are presented by different studies and standards available in the literature. First method relates electrical resistivity to chloride penetrability (mainly applicable to structures present in marine and industrial environments, where chloride ions are present) and the second method relates the electrical resistivity to the probability of corrosion of the reinforcement. Tables 1 and 2 show some resistivity values, associated with chlorides penetration and the probability of reinforcement corrosion, respectively, extracted from standards and scientific articles.

Recent studies relate electrical resistivity to chloride penetrability (Table 1). Balestra et al. (2020) present a classification proposal based on the analysis of the rate of variation of the chlorides concentration as a function of the surface electrical resistivity. Based on structures naturally degraded by marine environment, Balestra et al. (2020) defined the value of 5 kΩ cm as a reference for classifying a very high chloride penetration (consensus in the literature), while the intervals for the other

Table 1 Electrical resistivity values (kΩ.cm) associated to chloride penetration probability

Chloride penetration probability						References
Very high	High	Moderate	Low	Very low	Negligible	
–	<12	12–21	21–37	37–254	>254	AASHTO (2015)
–	<12	12–21	21–37	37–254	>254	FM 5–578 (2004)
–	<7	7–13	13–24.3	24.3–191	>191	FHWA-KS-14–15 (2015)
–	<16	16–28	28–50	50–343	>343	Ghosh and Quang (2015)
<5	5–7,5	7,5–15	15–35	>35	–	Sengul (2014)
≤5	(5, 9]	(9, 17]	(17, 41]	(41, 220]	>220	Balestra et al. (2020)

Table 2 Values of electrical resistivity (kΩ.cm) associated with corrosion probability

Chloride penetration probability						References
Very high	High	Moderate	Low	Very low	Negligible	
–	<10	10–50	50–100	–	>100	RILEM TC 154-EMC (2000)
–	<5	5–10	10–20	>20	–	CEB 192 (1989)
–	<6.7	6.7–11.7	11.7–20.6	20.6–141.1	>141.1	Azarsa and Gupta (2017)
–	<8	8–12	>12	–	–	(2004)
<10	10–50	–	50–100	–	>100	Polder (2001)
<15	15–20	20–55	55–80	–	>80	Balestra et al. (2019)

classes are established according to the possibility of chloride penetration associated with this limit value of electrical resistivity.

When relating electrical resistivity to the probability of reinforcement corrosion, it can be seen from Table 2 that the intervals defined by different references are wider than those presented in Table 1. In this sense, it should be emphasized that, while researches of (Balestra et al. 2019; Azarsa and Gupta 2017) are based on analyzes of concrete beams and structures in the field, respectively, the results of the other references presented in Table 2 are related to laboratory analyzes, using cylindrical specimens (100 mm in diameter and 200 mm high), under saturated conditions, at 28 days.

About electrical resistivity, the European standard UNE 83988-2 (AENOR 2014) highlights the need to take into account the geometry of the specimens and the distance between electrodes in the four-point electrical resistivity test, in order to minimize the possibility of errors in the results. Thus, for the determination of the resistivity of concrete this standard recommends the use of form factors (K_f) presented in Table 3.

Table 3 Form fator (k_f) of different specimens dimensions and probe spacing (AENOR 2014)

Specimen		k_f	
		Probes spacing (mm)	
Geometry	Dimensions (cm)	35	50
Cylindrical (Diameter x height)	30 × 15	0.714	0.606
	20 × 10	0.571	0.377
	15 × 7.5	0.384	–
Prismatic	4 × 4 × 16	0.172	–

3 Factors that Affect Electrical Resistivity of the Concrete

Concrete's electrical resistivity can be affected by different parameters, related to the physical characteristics of concrete and its constituent materials, as water/cement ratio, type of aggregates and cement used, mineral additions, presence of contaminants in the materials and saturation degree. The effects of these factors are presented and discussed.

3.1 Water/Cement Ratio, Saturation Degree and Presence of Contaminants

The water/cement ratio (w/c) is one of the principal parameters related to the porosity of concrete in the hardened state (and consequently to its mechanical strength). As high is this ratio, greater is the probability of the existence of a porous microstructure, with a large network of interconnected macro pores. In this context, a reduction in the w/c ratio, in addition to increasing the strength of the cementitious composite, also results in substantial increase of electrical resistivity (Andrade et al. 2009). Figure 3 shows an inversely proportional relationship between the electrical resistivity and the w/c ratio, where is demonstrated that by decreasing the w/c ratio by 25% results into an increase of concrete's electrical resistivity up to 90%.

In addition, it is widely recommended to determine the saturation degree of the concrete when electrical resistivity tests are performed, since the electric current is transported by ions that flow through the solution present in these pores, thus, as high is the internal moisture of the concrete, greater is the ion mobility, and consequently, lower is the electrical resistivity. Regarding ion mobility, Table 4 presents the electrical conductivity of the main ions normally found in the pores of concrete (Shi 2004). On the other hand, the effects of the degree of pore saturation can be analyzed in Fig. 4, where a significant increase in the electrical resistivity of the concrete can be observed as the saturation degree in its pores decreases.

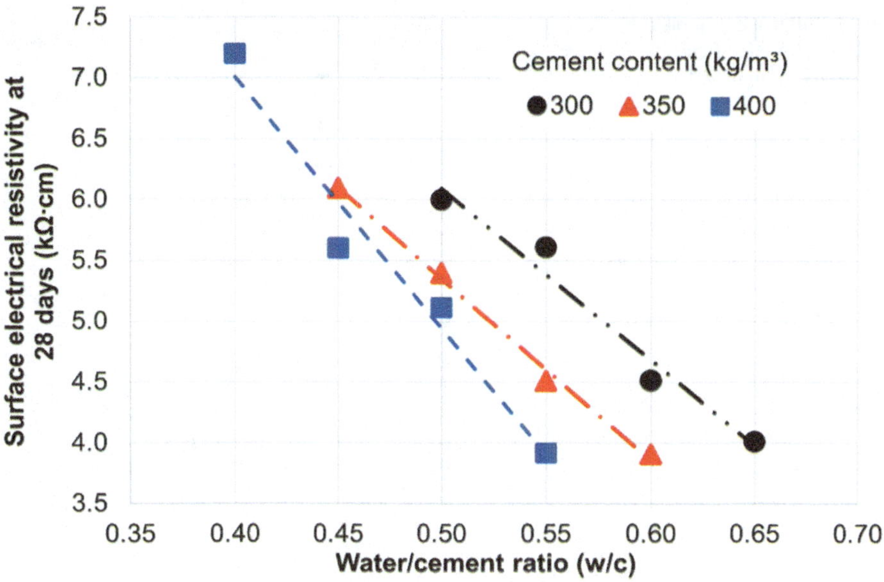

Fig. 3 Electrical resistivity and water/cement ratio relation (Hughes et al. 1985)

Table 4 Electrical conductivity of ions in concrete's pores (Shi 2004)

Ion	Na^+	K^+	Ca^{2+}	SO_4^{2-}	OH^-	Cl^-
λ_0 ($m^{-1}\Omega^{-1}$)	0.00501	0.00735	0.00595	0.00798	0.0198	0.00763

3.2 Aggregates, Cement Type and Additions

In concrete, it should be noted that as the cement hydration process occurs over time, microstructural changes occur in the pore network, with a refinement of the pore network, that contributes to increase the electrical resistivity. Thus, there are three possibilities of electric current conduction in the concrete: (1) Through the hydrated cement paste (Fig. 5a), (2) through the aggregate particles (Fig. 5b) or (3) through the paste and of aggregate particles (Fig. 5c) (Hou et al. 2017).

Generally, the electrical resistivity of the cement paste is lower than the aggregates, thus, an increase in particle size and aggregate content, in general leads to an increase in the electrical resistivity of the concrete, as shown in Fig. 6. The mineralogy of the aggregates also influences on resistivity, due to the different mineralogical elements that make up the rocks (Sengul 2014).

Regarding the type of cement used to produce concrete, the presence of supplementary cementitious materials (SCM), commonly observed in commercial cements, such as blast furnace slag and fly ash, result in benefits regarding electrical resistivity. At this point, the presence of these additions contributes, even in the early ages, to

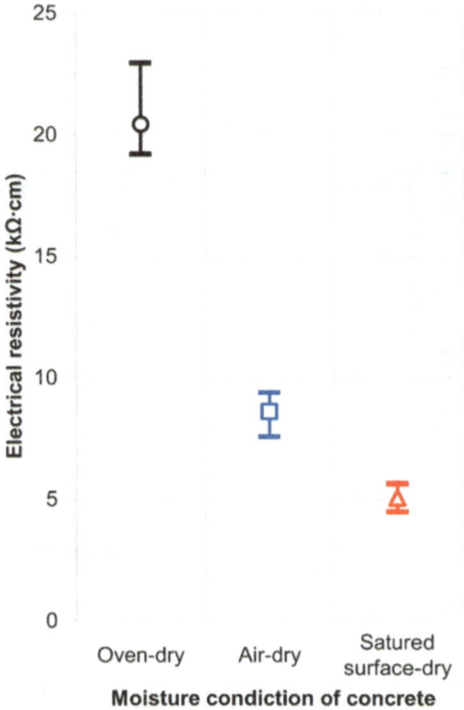

Fig. 4 Influence of moisture on concrete's electrical resistivity (Sengul 2014)

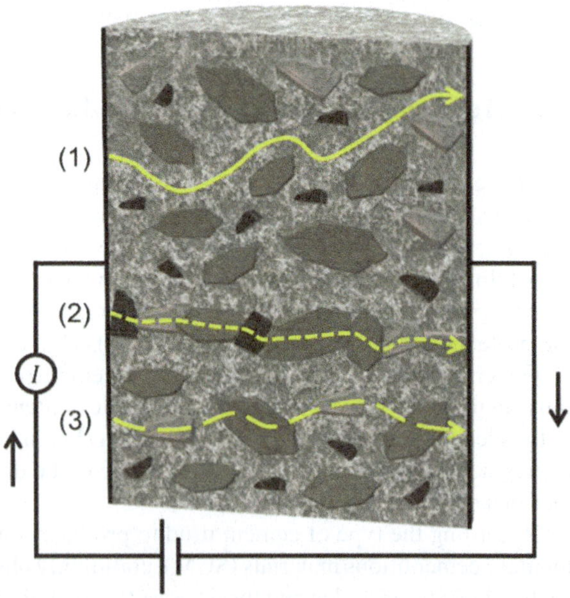

Fig. 5 Electrical flow through concrete (mesoscale) (Hou et al. 2017)

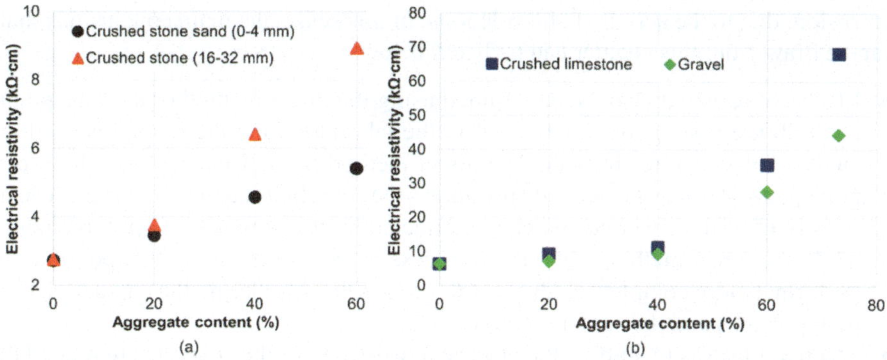

Fig. 6 Concrete's resistivity and aggregate type relation: **a** particles sizes (%) and **b** aggregate type (Sengul 2014)

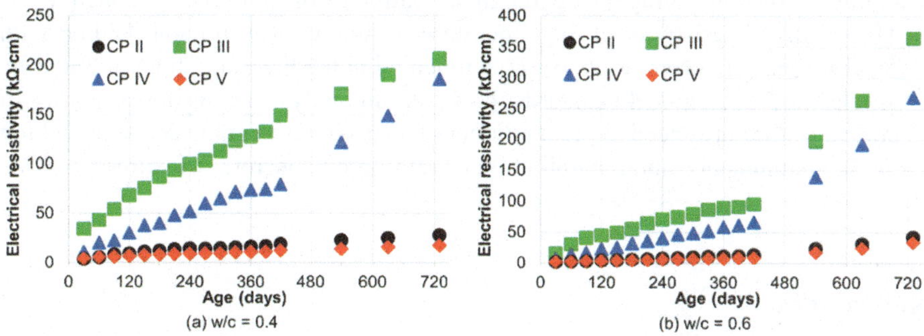

Fig. 7 Electrical resistivity and cement type relation for different water/cement ratio (Medeiros-Junior and Lima 2016)

increase the electrical resistivity of the concrete, due to a refinement of the porous microstructure of concrete, as observed for Brazilian's commercial cements in Fig. 7 (Medeiros-Junior and Lima 2016).

These differences in the electrical resistivity of cement types are explained by the SCM presents in cement. Cement type CP III has higher blast furnace slag content and CP IV has pozzolanic materials, while CP II has lower SCM quantities and CP V has no SCM.

4 Conclusions

The electrical resistivity of concrete is a non-destructive test, which evaluates the ion mobility through the pores of the concrete and, allows evaluating the durability and service life of concrete structures, mainly those subject to the reinforcement corrosion, due to the action of chloride ions. In this sense, the main conclusions that can be draw from this chapter can be listed below:

- Electrical resistivity can be determined using the direct method or the four-point method, the first is commonly used in the laboratory and the second is used in analysis of structures in field. The direct method has advantages over the four-point method, since this second only allows the superficial analysis of the electrical resistivity of the concrete and can be affected by the presence of reinforcement, while the direct method, although is necessary cores extraction, it is possible to perform a more complete analysis of the concrete, considering the characteristics of aggregates and the cement matrix.
- The main factors that affect the electrical resistivity of the concrete are related to its porous microstructure and the degree of pore saturation. In this case, emphasis should be given to the w/c ratio and cement consumption, as they control its porosity, as well as the environmental conditions of temperature and relative humidity, by controlling the degree of saturation of the pores of the concrete.
- The electrical resistivity of the concrete is a simple measurement method with results of easy interpretation, well documented in the literature, both in scientific articles and in technical standards, however, an analysis about the durability of concrete structures needs a great number of tests results to guide decision making on the maintenance and rehabilitation of reinforced concrete structures.

References

Akhtar A, Sarmah AK (2018) Construction and demolition waste generation and properties of recycled aggregate concrete: a global perspective. J Clean Prod 186:262–281

American Society for Testing And Materials (2012) ASTM C 1760: Standard Test Method for Bulk Electrical Conductivity of Hardened Concrete. United States

American Association of State Highway and Transportation Officials. AASHTO T 358–15 (2015) Standard Method of Test for Surface Resistivity Indication of Concrete's Ability to Resist Chloride Ion Penetration. Washington, pp 1–10

Andrade C, D'andrea R, Castillo A, Castellote M (2009) The use of electrical resistivity as NDT method for the specification of the durability of reinforced concrete. NDTCE'09, Non-Destructive Testing in Civil Engineering Nantes, France

ASSOCIAÇÃO ESPANHOLA DE NORMALIZAÇÃO E CERTIFICAÇÃO (AENOR) (2008) UNE 83988-1 Durabilidad del hormigón. Métodos de ensayo. Determinación de la resistividad eléctrica. Parte 1: Método directo

ASSOCIAÇÃO ESPANHOLA DE NORMALIZAÇÃO E CERTIFICAÇÃO (AENOR) UNE 83988-2 Durabilidad del hormigón: métodos de ensayo: determinación de la resistividad eléctrica. Método de las cuatro puntas o de Wenner, Part 2: Método de las cuatro puntas o de Wenner (2014)

Azarsa P, Gupta R (2017) Electrical resistivity of concrete for durability evaluation: a review. advances in materials science and engineering, pp 0–302017

Balestra CET, Nakano AY, Savaris G, Medeiros-Junior RA (2019) Reinforcement corrosion risk of marine concrete structures evaluated through electrical resistivity: Proposal of parameters based on field structures. Ocean Eng 187:106167

Balestra CET, Reichert TA, Pansera WA, Savaris G (2020) Evaluation of chloride ion penetration through concrete surface electrical resistivity of field naturally degraded structures present in marine environment. Constr Building Mater 230, p 1, 116979

Chen CT, Chang JJ, Yeih W (2014) The effects of specimen parameters on the resistivity of concrete. Constr Build Mater 71:35–43

COMITÉ EURO-INTERNATIONAL DU BÉTON—CEB. Diagnosis and Assessment of Concrete Structures—State-of-Art Report. CEB Bulletin d'Information, n.192, 1989

Federal Department of Transportation. FM 5-578 (2004) florida method of test for concrete resistivity as an electrical indicator of its permeability. FDOT, Flórida, USA, 4 p

Ghosh P, Quang T (2015) Influence of parameters on surface resistivity of concrete. Cement Concr Compos 62:134–145

Hughes BP, Soleit AKO, Brierley RW (1985) New technique for determining the electrical resistivity of concrete. Magazine Concrete Res 37(133):243–248

Hou TC, Nguyen VK, Su YM, Chen YR, Chen PJ (2017) Effects of coarse aggregates on the electrical resistivity of Portland cement concrete. Constr Build Mater 133:397–408

Kansas Department of Transportation Bureau of Research FHWA-KS-14-15 (2015) Surface resistivity as an alternative for rapid chloride permeability test of hardened concrete. Kansas Department of Transportation Bureau of Research. Kansas, USA 39 p

Mehta PK, Monteiro PJM (2006) Concrete: microstructure, properties and materials. 3th edition. Mc-Graw Hill. New York, 674 p

Medeiros-Junior RA, Lima MG (2016) Electrical resistivity of unsaturated concrete using different types of cement. Constr Build Mater 107:11–16

Neville AM, Brooks JJ (2010) Concrete technology. 2st ed. Pearson. London, 448 p

Nguyen AQ, Klyz G, Derby F, Balayssac J-P (2017) Evaluation of water content gradient using a new configuration of linear array four-point probe for electrical resistivity measurement. Cement Concrete Compos 83:308–322

Nguyen AQ, Klyz G, Derby F, Balayssac J-P (2018) Assessment of the electrochemical state of steel in water saturated concrete by resistivity measurement. Constr Building Mater 171:455–466

Polder RB (2001) Test methods for on site measurement of resistivity of concrete—a RILEM TC—154. Constr Building Mater 15:125–131

Rilem TC 154-EMC (2000) Electrochemical techniques for measuring metallic corrosion. Mater Struct/Matériaux et Constr 33:603–611

Sengul O (2014) Use of electrical resistivity as an indicator for durability. Constr Building Mater 73:434–441

Smith KM, Schokker AJ, Tikalsky PJ (2004) Performance of supplementary cementitious materials in concrete resistivity and corrosion monitoring evaluations. ACI Mater J 10(5):385–390

Shi C (2004) Effect of mixing proportions of concrete on its electrical conductivity and the rapid chloride permeability test (ASTM C1202 or ASSHTO T277) results. Cement Concrete Res 34(3):537-545

Rehabilitation Technologies

W. Mazer, E. O. Ferreira, M. T. Q. Simões da Silva, and R. Basso Tokarski

Abstract Several pathological manifestations existing in the buildings lead to a decrease in the durability and useful life of the building. These problems require an adequate correction to be made to restore the building's conditions of use. To make these corrections, there are several techniques and materials that can be used, including the reconstitution of concrete, crack injection techniques, surface protection, electrochemical techniques and structural reinforcement, some of which are addressed in this work.

Keywords Structural rehabilitation · Repair systems · Electrochemical techniques · Surface protection · Injection systems

1 Introduction

Aging and environmental aggressiveness cause the buildings to wear out, generating the need for maintenance, whether preventive or corrective. In order for this maintenance to restore the useful life of the buildings, it was necessary to develop specific techniques and materials for the repair and rehabilitation services.

Among the available techniques, one of the most used ones refers to concrete repair systems, which consist of the recovery of concrete failures using materials such as high-performance or projected concrete and structural polymeric mortars, with the aim of recomposing the concrete section.

W. Mazer (✉)
University of Technology-Paraná, Curitiba, Brazil
e-mail: mazer.utfpr@gmail.com

E. O. Ferreira · M. T. Q. Simões da Silva · R. Basso Tokarski
Federal University of Technology-Paraná, Curitiba, Brazil
e-mail: elizamaryotto@gmail.com

M. T. Q. Simões da Silva
e-mail: marlytsilva@utfpr.edu.br

R. Basso Tokarski
e-mail: rosangela.tokarski@ifsc.edu.br

J. M. P. Q. Delgado (ed.), *Durability of Concrete Structures*, Building Pathology and Rehabilitation 16, https://doi.org/10.1007/978-3-030-62825-3_5

The crack injection technique is used to fill cracks and the materials used vary according to the purpose, whether structural or waterproof, with rigid or flexible materials.

In order to prevent the appearance of damage and also to protect the structure after the recovery process, surface protection techniques such as paintings, mortar coverings, product impregnation or even waterproofing must be employed.

Electrochemical techniques are used to recover structures with carbonation or chloride penetration processes, and cathodic protection can also be used to prevent the appearance of problems.

Finally, it may be necessary to reinforce the structure, and in this case, among the various existing techniques, reinforcement with polymers reinforced with carbon fibers can be used, which has been used with great success, including improving the performance of structures in places subject to earthquakes.

2 Concrete Repair Systems

2.1 Reconstitution of Concreting Failures

Concreting failures originate during the concreting process caused by improper casting or thickening of the concrete, they can also be caused by excess reinforcement in a section which can cause the segregation of the concrete (this segregation can also be caused by the use of incorrect diameter of the aggregates), causing failures in the structure, which is one of the most recurrent pathological manifestations in concrete (Helene 1988).

Moreno et al. (2007), advises that the reconstitution of these failures should resume or restore the initial states of internal efforts and the fractions of performance and service of the structure. The choice of materials and the correction technique to be employed depends on the diagnosis of the problem, the characteristics of the region to be corrected and the operating requirements of the concrete element, object of the correction (Helene 1988).

Cusson and Mailvaganam (1996) point out that in order to obtain a durable repair when they are more superficial it is essential that the properties of the repair material and the substrate be correctly combined. In addition, perfect adhesion between the substrate and repair material is required (Abu-Tair et al. 2000). This should ensure that the repair material could withstand the stresses resulting from the loading and volume changes of the structure, in a specific environment, without experiencing signs of early deterioration, cracking or detachment from the substrate.

The materials used in the reconstruction of defects must present mechanical and physical properties equal or superior to materials to be reconstituted, as well as having low shrinkage and permeability, good durability, strength, workability, adhesion to the substrate and be easily applied. Mortars and shotcrete are among the main materials used to reconstitute concrete failures.

Cusson and Mailvaganam (1996) propose three generic groups for classifying repair mortars for concrete: cement mortars, cement mortars modified by polymers and resin mortars. The combinations of the types of binder and polymeric or inorganic additions determine a reasonable range of properties and possibilities for the formulation of repair mortars (Moreno et al. 2007).

The importance of choosing the material for reconstitution of the faults stands out in the compatibility of the mechanical efforts of the materials so that there are no cracks between the interface of the repaired concrete substrate and the repair carried out on the structure, which can also cause a loss of the load capacity of the structure.

Cusson and Mailvaganam (1996) cite that the large number of repair materials available on the market, with a wide variation in mechanical properties, makes the proper selection of them an extremely judicious work, especially because environmental and environmental conditions must be taken into account the adverse load.

2.2 *Shotcrete*

Shotcrete is a concrete that is transported by a hose, from the spraying equipment to a nozzle, which by means of compressed air projects it at great speed against a surface.

According ACI-506 (1994), shotcrete is a "mortar or concrete pneumatically shot at high speed on a surface". The literature presents more definitions for shotcrete, but all of them emphasize the high speed of spraying the concrete as a condition to obtain the desired final result.

According Wood et al. (1990), the key to the durability of a structural repair is to ensure the continuous transfer of load and stresses between the concrete and the repair material, without the occurrence of cracks in the interface that cause loss of adhesion and detachment of the repair mortar.

For this to occur, the materials to be used must have characteristics such as: modulus of elasticity similar to that of the substrate, low permeability, minimum shrinkage, coefficient of thermal expansion similar to that of concrete, ensuring good adhesion to steel and concrete, having compatible properties.

The versatility that shotcrete presents, makes it an important product in civil construction. Reconstituting gaps or concreting nests is one of the uses of this product, some of the advantages of shotcrete are: The rate of resistance development (compression and traction), short-term dimensional stability and bond strength to damaged surfaces, that is, high adhesion power of shotcrete to the substrate to be repaired.

The shotcrete also has a high impermeability preventing the penetration of moisture and gases through reconstitution, thus preserving the integrity of the repaired structure.

The execution of shotcrete can be dry or wet, and in both cases there is a problem of reflection that consists of the loss of material that is not adhered to the surface.

Tadolini et al. (2018) developed a shotcrete with a stabilizing capacity that can be used to restore the structural integrity of the damaged structure resulting in structural strength within minutes, allowing access to damaged structures within minutes of applying the mixture formulation. The product consists of dry mixes from a single bag, which facilitates rapid transport to any location, and can be used, in simple equipment.

2.3 Concrete Replacement

The replacement of concrete is usually used in the treatment of concrete failures or breakdowns. The techniques used in the treatment of failures vary according to the magnitude of the problem. When the fault is superficial, reaching only the reinforcement covering region, the restoration can be carried out by cleaning the region properly, applying an adhesion bridge and recomposing the section of the piece with a structural mortar based on epoxy or polymeric.

For the recovery of large areas and more serious problems, special care must be taken. First, an adequate shoring for the part must be studied. The substrate must be prepared considering that the concrete around the fault may be damaged, then the substrate is cleaned, the adhesion bridge is applied and the part section is reconstituted. A concrete or micro-concrete can be used with a compressive strength higher than the concrete of the piece, or even grout or epoxy mortar. The material used for the repair must not shrink.

Before proceeding with the recovery of concrete attacked by disintegration, it is of fundamental importance to carry out an analysis of the concrete and the environment to which it is exposed, in order to determine the causes that cause the pathology. In general, the repair consists of replacing the attacked region, carrying out an adequate shoring of the element, preparing and cleaning the substrate, reconstituting the section with an appropriate concrete and the proper protection of the surface of the part.

In these cases, the repair concrete must be a HPC with a very low void index and adequate resistance to the aggressive medium. The surface protection can be done with epoxy resin or water-repellent silicone.

The success in the reconstitution of the failures is in the resistance between the repair and the interface of the substrate to be repaired, the way in which the concrete will be replaced and the preparation of the surface that will receive the new concrete is fundamental for achieving success in the reconstruction of the failure. The preparation of the substrate consists of removing the damaged material, usually done through mechanical scarification, made with a pneumatic or electric hammer, and requires special care in its application in order not to damage the structure. This process has the objective of opening the pores of the substrate to generate roughness and sufficient cleaning to give the necessary adhesion to the substrate and reconstitution material interface.

Júlio et al. (2004), demonstrated in their study the result of using different techniques for preparing the substrate with superficial roughness in the bond strength. The techniques studied were manual brushing with a metal bristle brush, sandblasting and partial thinning using a needle gun, using the pull-off and slant shear test as the adhesion measure. The authors observed that sandblasting resulted in a greater adhesion between substrate and repair material for the two tests to which they were subjected, and that the pre-wetting of the substrate surface does not significantly interfere in the resulting adhesion resistance.

Abu-Tair et al. (2000) affirm that hand-chiseled surfaces increase, on average, 25% the adhesion resistance in relation to surfaces prepared with a pistol. The same authors also conclude that around 10% of the surfaces tested with a pistol failed to bond, which may suggest that the action of the pistol caused damage to the concrete of the substrate and caused failures just below the treated surface.

If there are exposed armatures in the region to be recovered, it is also necessary to prepare them for perfect adherence. This preparation can be carried out by; sanding or manual or mechanical brushing, jets of water under pressure or jet of compressed air, ensuring the elimination of any existing corrosion.

Regarding the bonding of the substrate with the new concrete, Elbarky et al. (2016) concluded that increasing the roughness of the substrate's concrete surface by hand chisel is considerably more effective than grinding. The adhesion resistance obtained for samples treated by manual chiseling was, on average, 3.19 times greater than the adhesion resistance obtained for samples treated by grinding; resinous agents used between the substrate concrete and new concrete do not increase the bond strength between the materials used.

3 Injection Systems

The selection of the material to be used in filling cracks depends on factors such as thickness, humidity and daily movement of the crack and the selected material must have one of the following functions (Raupach and Büttner 2014):

- Filling with transmission of efforts;
- Ductile filling;
- Swelling fitted filling of cracks

The main materials used are; epoxy resins, polyurethane resins, acrylic gels and cementitious materials, and the properties of each material and compatibility with the substrate must be observed before choosing the most suitable material. According to Sika (2020) there are 3 important factors in the effectiveness of an injection system, namely the material, the equipment and the application method, in addition to the training of the workforce. In the material item, viscosity, flexibility and behavior when in contact with water must be observed. For the equipment, in addition to the type of pump suitable for the material, the correct nozzle must also be chosen, while the application method involves prior knowledge and experience in the injection applications.

3.1 Epoxy Resins

Epoxy resins are considered a rigid repair system so they should only be used in passive cracks and have the purpose of returning monotheism to the structural element. Due to their low viscosity (150 to 400 m Pa. s) they can be used in cracks with an opening greater than 0.1 mm and this characteristic also helps in load transfer. In general, epoxy resins have great resistance to compression and traction, in addition to excellent adhesion to concrete, but the presence of moisture can decrease both adhesion and strength, for this reason it is recommended that its application be made on dry substrate

3.2 Polyurethane Materials

Polyurethane-based materials for injection into cracks can be one-component or two-component, and can also be in the form of foam or gel. The polyurethane foam products are used for a primary injection in a place with a large flow of water, reducing this flow and allowing the injection of products in the form of gel to stop the leak. Normally polyurethane products have low viscosity, are hydrophobic and flexible, being the only products indicated for the case of fissures that show movement over time.

3.3 Acrylic Resins

Acrylic resins are low viscosity, flexible and hydrophilic materials, used for filling non-structural cracks and sealing. They are usually used for soil stabilization and for forming waterproofing membranes in concrete.

3.4 Cement Materials

The injection with cementitious materials can be done with cement cream or with cement suspension. The difference between the methods occurs by the size of the grains, where the cement cream uses grain with a diameter of up to 100 cm, while the suspension uses smaller grains, which according to Raupach and Büttner (2014) 95% of the material has a maximum diameter of 16 cm the specific surface of 16,000 cm^3/g. Because it is a rigid material, it is used to restore the monotheism of the structural element and in the transmission of loads, however it cannot be used in cracks that show movement over time. To improve its properties and assist in the injection process, modified polymers are added to its composition.

3.5 *Equipment*

Basically, two types of equipment are available for the execution of the injection services, the single-component pumps and the two-component pumps. Single-component pumps can be manual, pneumatic or electric with diaphragm and are used to apply small and medium volumes of cementitious materials or single-component resins. In two-component pumps, the mixing between products occurs in the pumping process, with specific pumps for each type of resin depending on the viscosity, the mixing ratio, the time of action among other characteristics, being indicated for medium and large volumes of injection of bicomponent products.

In addition to the injection pumps, it is necessary to use injection nozzles for the execution of the service, and there are injection nozzles that are installed in holes previously made in the structure and can be used in high or low-pressure systems and injection nozzles. Surface installation, when it is not possible to pre-drill the structure, in which case only a low pressure system can be used.

3.6 *Application Systems*

The main use of the injection system is to fill in cracks and according to the injection nozzle used, the execution system changes. For mechanically installed injection nozzles, which are placed in holes in the part, holes should be drilled at half the thickness of the part when the injection is made only on one side of the structure or at a distance of 25% of the thickness of the part when the injection is carried out on both sides of the part, inclined at 45° in relation to the surface and the holes must be alternated in relation to the crack side, as shown in Fig. 1 and the diameter of the

Fig. 1 Detail of the installation of injection nozzles (*Source* Techniques 2020)

hole must be equal to the diameter of the injection nozzle added 2 mm. For surface injection nozzles, the distance between them should be 90% of the thickness of the part for injection on only one side or 50% of the thickness for injection on both sides of the part.

When leaks occur over large areas, a recovery method is through the curtain injection system behind the structure. For this purpose, mechanical installation nozzles are used and the holes must be spaced between 30 cm and 50 cm across the entire area to be waterproofed.

4 Surface Protection Systems and Barriers

The atmosphere presents agents that enhance the degradation of concrete and the corrosion of reinforcement, such as oxygen, organic peroxides, acidic materials such as carbon dioxide, salts, such as sodium and alkali chloride (Sherwood 1966). One of the techniques to protect structures from environmental aggressiveness is surface protection; the main methods used being coating with paints, mortar coatings, impregnations and waterproofing.

4.1 Paint Coating

One of the protection methods used to prevent corrosion from these agents is the application of paints on surfaces; this is due to the fact that it is a low-cost material, with easy application. The principle of protection of this type of material is related to the permeation of electrolytes that occurs (Bethencourt et al. 2003).

The chemical bonds between the compounds present in the environment and the products of the paints, the degree of curing of the material and the crosslinking agent are the factors that affect the chemical and mechanical resistance of the coating, being its adhesion in the structure or the ability to resist the spraying of salt and the weather (Almeida et al. 2006).

The main paints applied for protection are water-based paints; powder paints and high-strength paints. These materials act as physical and cathodic protection barriers for the structure. Epoxy-based paints with a zinc-rich primer have a greater protection capacity compared to others, since they have lower water absorption rates (Almeida et al. 2006).

When painting, epoxy or polyurethane, is applied on the reinforcement for protection against corrosion, the adhesion between the reinforcement and the concrete must be checked, so as not to alter the structural performance.

4.2 Mortar Coating

The protection with mortar coating generates continuous protection on the surface of the structure, eliminating the penetration of aggressive agents. The application of cement-based mortars is a methodology used because it is a material with a low shrinkage factor, being applied in large parts of reinforced concrete recovery works. There are two main variations, self-leveling mortars—used on horizontal surfaces because they have great fluidity—and thixotropic mortars—applied at different angles in relation to the horizontal because they do not flow, they are mortars that have the addition of polymers (oliveira and Alves 2018).

Epoxy-based mortars—resulting from the combination of epochloridine and biphenol—are characterized by the use of catalysts for polymerization to occur, the proportions of the materials used interfere in the final properties of the material and because the binders present in the mixture are insoluble in water its use is common to combat corrosion of the reinforcement bars. The shrinkage of this material is almost null during curing, but its expansion coefficient is higher than that of concrete, not being used in places with high temperatures. (Souza and Ripper 1998).

Geopolymer based mortars are formed by mixing common aggregates with the addition of metakaolin, calcium hydroxide and alkaline silicate solution. The proportion of metakaolin added is between 5 and 10%, this generates protection against chloride ions and the carbonation of reinforced concrete structures (Vasconcelos et al. 2011). According to Curcio et al. (1998), replacing 15% of cement with metakaolin generates an increase in pozzolanic activity in mortars and improves final strengths, decreasing the material's porosity.

Within the research that uses mortars as material for the recovery of structures, Feng-qing et al. (2011) developed a cementitious material made with steel slag, granulated blast furnace slag, fly ash and the catalyst, in addition to replacing part of the quartz sand by iron waste, the presence of these materials generates a mortar with better mechanical properties and greater water resistance.

4.3 Impregnations

According to Raupach and Büttner (2014), impregnation systems are used to decrease surface porosity, limiting the penetration of liquids and gases and increasing the surface resistance of concrete, through the formation of a discontinuous film that fills the surface pores.

Scarfato et al. (2011) carried out studies of surface treatments using Nano composites with organophilic clay based on commercial resins in order to develop a barrier to water—liquid and vapor—and attacks of sulfates. Using resins based on vinylidene fluoride and siloxane with the addition of up to 6% of organophilic clay, it is possible to identify the Nano filler behavior of the clay, closing the pores of the concrete substrate, improving mechanical and barrier resistance properties, as well as better behavior in relation to thermal stability.

It is possible to carry out the surface treatment through the application of hydrophobic products, which generate a water-repellent surface but do not fill the pores of the material.

Also in impregnation, surface-hardening products can be used, which are based on sodium silicate or magnesium fluorosilicate, which increase the abrasion resistance of the surface.

When concretes are treated with silicate-based products mixed in an acrylic resin, it is possible to identify improvements in abrasion resistance and waterproofing without the need to increase the surface roughness (Baltazar et al. 2014).

4.4 Waterproofing

When concrete structures are subject to the action of water, their protection can be done through waterproofing systems, and the choice of the appropriate system depends on several factors, among which the movement of the structure must be considered. For structures that are subject to temperature variation or expansion joints, flexible-waterproofing systems must be used and in the case of structures that do not undergo movement, a rigid waterproofing system can be used.

The most widely used flexible systems are asphalt blankets. For the application of this system, it is necessary to prepare the substrate, eliminating irregularities and applying paint with asphalt emulsion. Then the hot blanket is applied and after the hydrostatic test mechanical protection is performed.

Rigid systems are basically composed of cementitious materials modified with polymers. They are usually applied through painting with 4 crossed coats to avoid failures in the application.

Wang et al. (2020) developed research in relation to hydrophobic mortars, using direct dispersion of pre-modified cement and sand particles in aqueous solution of polyvinyl alcohol surfactant, this makes it a material with low surface energy, being hydrophobic both to water when aggressive agents such as hydrochloric acid with $pH = 5$ and sodium hydroxide solution.

5 Electrochemical Techniques

Electrochemical techniques for protecting structures consist of applying a DC to the structure to prevent or paralyze reinforcement corrosion. The main techniques used are cathodic protection, electrochemical realization and electrochemical chloride extraction.

5.1 Cathodic Protection

Cathodic protection is defined as the reduction or elimination of corrosion due to the transformation of armatures into cathodes, due to the imposition of galvanic or printed current.

The cathodic protection principle is based on taking the corrosion potential of the part to be protected to values corresponding to the material immunity, according to the Pourbaix diagram, illustrated in Fig. 2.

The concept of cathodic protection is closely related to the potential diagrams—Pourbaix pH, and the electrochemical polarization curves. Among the existing cathodic protection methods, sacrificial anodes and the printed current can be mentioned.

The sacrificial anode method consists of protection by connecting the structure to a more active metal immersed in an electrolyte. The result is a galvanic cell in which the active metal works as an anode, promoting a flow of electrons to the structure, which becomes cathodic. In this case, the anode is destroyed progressively and is called a sacrificial anode.

This protection is based on the electrochemical series of metals where the noblest metals do not corrode and the less noble metal suffer corrosion. When a less electronegative metal (more positive potential) is in electrical contact with a more electronegative metal (more negative potential), and with the presence of moisture and oxygen and due to the potential difference between the metals, a current flow will appear, called galvanic effect, and in this way steel will be protected through cathodic protection (Davies and Broomfield 2014).

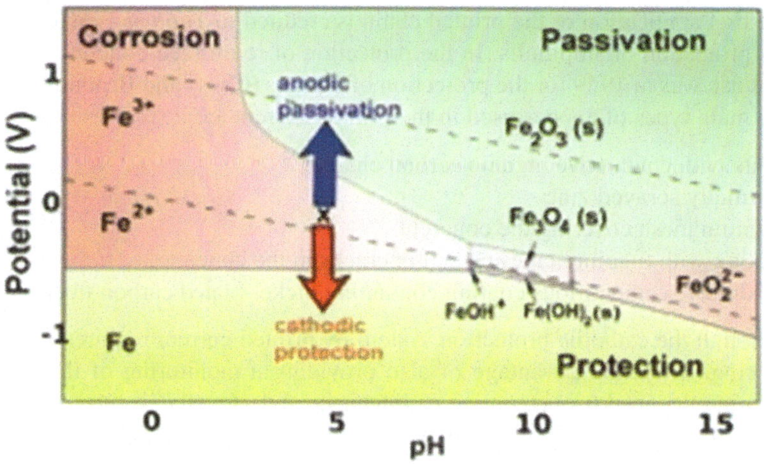

Fig. 2 Pourbaix diagram for iron (*Source* https://steemit.com/corrosion/@rahim.rahman/corrosion-pourbaix-diagram)

When additional positively charged anodes are introduced into a reinforced concrete structure, it is possible to generate an electric current in the steel, which is negatively charged, in order to control corrosion. This procedure prevents the iron metal from oxidizing to form the iron ion and release electrons, as indicated in Eq. (1) (Davies and Broomfield 2014).

$$Fe = Fe^{2+} + 2e^{-} \tag{1}$$

At the same time, next to the reinforcement, a cathodic reaction will occur, forming hydroxyl ions that help to re-alkalinize the concrete. If the structure is carbonated, this procedure helps in the restoration and maintenance of the steel passivation layer and if the structure is attacked by chloride ions, the negative charge in the armature prevents the chloride ions from approaching, which are attracted to the anode (Davies and Broomfield 2014).

The first sacrificial anodes used were made of iron-silicon used to protect buried pipes. In the 1990s, titanium mesh anodes encased in metal oxides were developed and, among other applications, they were also used in bridge protection (Chess and Broomfield 2014).

The printed current protection system consists of installing an anode bed on the concrete surface, protecting this bed with a coating layer, generally cementitious, and using an external source (rectifier) applying a cathodic current (positive pole connected anode and negative to the armature) in order to make the armature cathodic.

There is a big difference with the previous method, since in that the polarization occurs spontaneously by the contact between the two metals. In the printed current process, a generator or rectifier supplies the current externally. The current rate inserted in the structure varies from 0.2 to 2.0 mA/m^2 of steel area to be protected.

The first application of the printed chain is credited to Thomas Edison, in 1890, for the protection of ship hulls. In the protection of reinforced concrete structures, the first use was in 1959 for the protection of bridges (Chess and Broomfield 2014).

The main types of anodes used in the printed current are:

- Paints with conductive organic carbon charge.
- Thermally sprayed zinc.
- Titanium mesh covering the concrete.
- Coating with titanium tapes placed in cracks in the concrete.
- Conductive cementitious coating containing nickel-plated carbon fibers.

Although the cathodic protection system by printed current requires continuous monitoring, it has the advantage of also providing a monitoring of the structure, anticipating the need for preventive maintenance and, if well designed, presents the least interference in the structure (Broomfield and Mcdonald 2003).

5.2 *Electrochemical Re-alkalinisation*

According Mietz (1995) electrochemical re-alkalization is a non-destructive method of recovering structures that consist of restoring the pH of the concrete to provide the passivation of the reinforcement.

This technique is based on the passage of an electric current between an external metallic mesh that functions as an anode and the structure's reinforcement. An alkaline solution is also used as an electrolyte covering the metal mesh and the concrete part (Yeih and Chang 2005).

For the application of this technique, it is necessary to wrap the affected area with a geotextile blanket saturated with the alkaline solution and the metallic mesh. After making the electrical connection between the anode and the armature, an electric charge is applied with a current that must be between 0.5 to 1 A/m^2 with two poles, the negative pole being connected to the armature (forming a cathode) and the positive pole is interconnected in the mesh (forming an anode). During this process, oxygen is reduced and hydroxyl ions (OH-) are formed next to the armature, which makes it possible to pacify it (Goyal et al. 2018). In addition, the penetration of the alkaline solution helps to restore the pH of the concrete.

Many factors influence the effectiveness of the recovery method, in this sense; Lachovicz (2020) evaluated the use of aluminum mesh, galvanized steel and stainless steel as anodes of the system for a period of 7 days. The aluminum mesh, despite not showing corrosion, suffered damage, but the complete realization of the material was verified, as shown in Fig. 3. The galvanized steel mesh suffered corrosion and the concrete was not realized, as shown in Fig. 4. The stainless steel mesh suffered

Fig. 3 Result of realcalinization with the aluminum mesh. **a** State of the mesh at the end of the process. **b** Realkalinized specimen (*Source* Lachovicz 2020)

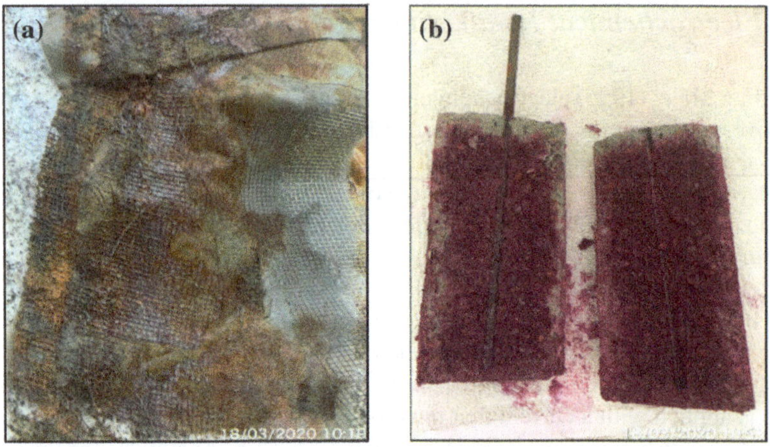

Fig. 4 Result of re-calcination with the galvanized steel mesh. **a** State of the mesh at the end of the process. **b** Realkalinized specimen (*Source* Lachovicz 2020)

damage at the points of contact with the rectifier source and with the armature, however the re-calcination was also complete, and the steel was completely clean after the process, as shown in Fig. 5.

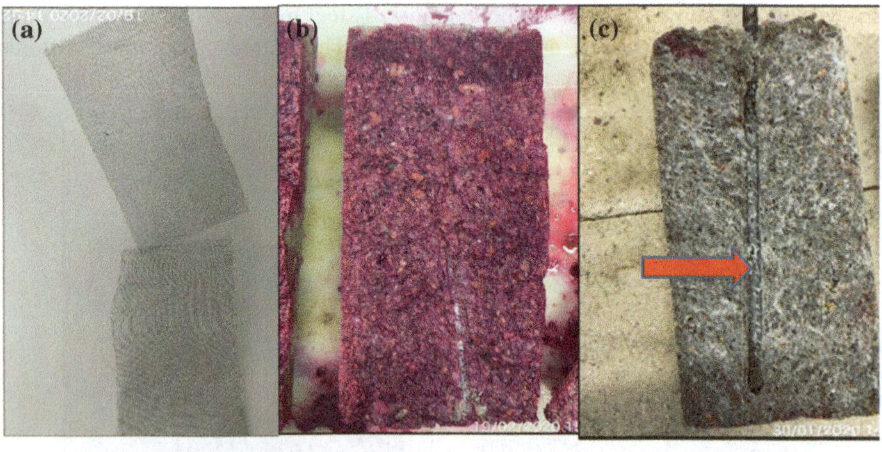

Fig. 5 Result of realcalinization with the stainless steel mesh. **a** State of the mesh at the end of the process. **b** Realkalinized specimen. **c** Reinforcement without corrosion (*Source* Lachovicz 2020)

5.3 *Electrochemical Chloride Extraction*

It is a method of removing Chloride ions present in reinforced concrete structures through the use of a temporary electrochemical current. It consists of the application of an electric field between an anode, fixed externally to the structure, and the armature, which will work as a cathode, in the presence of an electrolyte, usually sodium hydroxide.

According to McDonald and Broomfield (2003) this technique uses a current of 1A/m^2 for a period of up to 8 weeks, being possible to remove 50% to 90% of the chloride ions in the reinforced concrete structures.

Dyer (2014) mentions that when a potential difference is established between the electrodes installed on the concrete surface and the reinforcement, the chloride ions migrate to the surface. The principle of this method is similar to cathodic protection by printed current, in which the formation of hydroxyl ions occurs next to the armatures, repelling chloride ions, which at the same time are attracted by the external anode.

Schneck (2011) mentions that at the beginning of the treatment, the movement of chloride ions is greater, and decreases over time while an increase in hydroxyl ions occurs.

The efficiency of the method depends on factors such as the water/cement ratio, the concentration of chloride ions, the density of the applied current, the covering of the reinforcement, among others. However, as well as electrochemical realcalinization, this method has some side effects, such as increased porosity of concrete, reduced adhesion between concrete and steel and the possibility of RAA occurrence.

The Table 1 presents a resume of the electroquimical technique presented.

Table 1 Review of the eletroquimical techniques

Electrochemical technique	Cathodic protection	Electrochemical re-alkalization	Electrochemical chloride extraction
Protection objective	Reinforcement immunization	Increase the pH	Chloride extraction
Applicability	Any structure	Carbonated structure	Structures with Chlorides
Applied current	10 mA/m^2	0.2 to 2 A/m^2	0.1 to 2 A/m^2
Application time	Permanent	1–2 weeks	4–10 weeks
Benefits	Unique effective against initiated corrosion	Without removing the concrete	Without removing the concrete
Disadvantages	Monitoring	Difficulty in field application	Difficulty in field application

6 Reinforcement Systems with Carbon Fibers

In some situations, recovery alone is not enough to rebuild the structure, and it is necessary to use structural reinforcement techniques. Among the various existing techniques, structural reinforcement with carbon fiber reinforced composites stands out.

It is possible to define composite as a multi-phase combination material, that is, of different molecular structures with different properties (Callister and Rethwisch 2012; Poletto 2017). It consists of two phases: the matrix phase and the dispersed phase, as illustrated in Fig. 6. The matrix phase can be metallic, organic or ceramic and the dispersed phase is formed by carbon fibers, glass fibers, aramids or natural fibers properties (Callister and Rethwisch 2012; Gomes 2015).

Carbon fiber composites are replacing metal-based structural materials thus providing lighter, stronger and more durable solutions for use in industry and construction (Koumoulos et al. 2019).

In civil construction, carbon fibers in the form of composites are used to repair reinforced or prestressed concrete structures. They have resistance to corrosion, provide increased resistance to shearing and bending of beams and slabs and increased containment of columns, in addition to increased stiffness of concrete structures (Nanni 2000).

The ease of installation as well as its resistance to the environment and the properties of carbon fibers has accelerated the acceptance of the use of carbon fiber composites by civil engineers worldwide (Parmar et al. 2018). According to Lazo and Rinchon (2018) composites reinforced with carbon fibers are also presented as an alternative for reinforcement against earthquakes.

Among the application techniques of the structural reinforcement system with CFRP, the most common use is the application with CFRP sheets or blankets adhered externally (Externally Bonded Reinforcement—EBR) as illustrated in Fig. 7, however there is also the application of sheets or CFRP bars inserted in the concrete cover thickness (Near Surface Mounted—NSM) (ACI 2008) shown in Fig. 8.

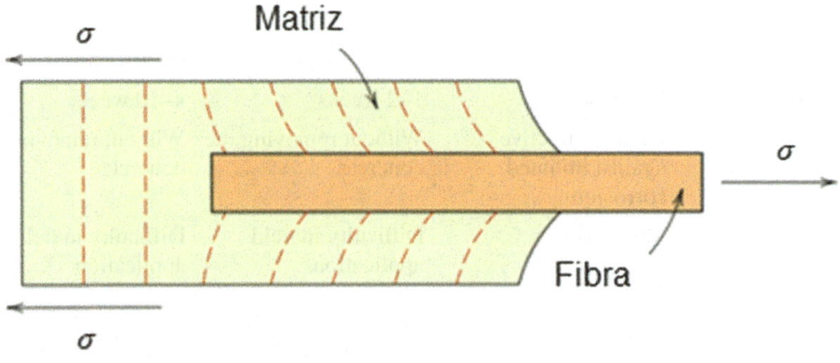

Fig. 6 Deformation pattern in the Matrix Phase around a Scattered Phase (Fiber) that is subjected to the application of a tensile load (*Source* Callister et al. 2012)

Fig. 7 Application of the EBR system

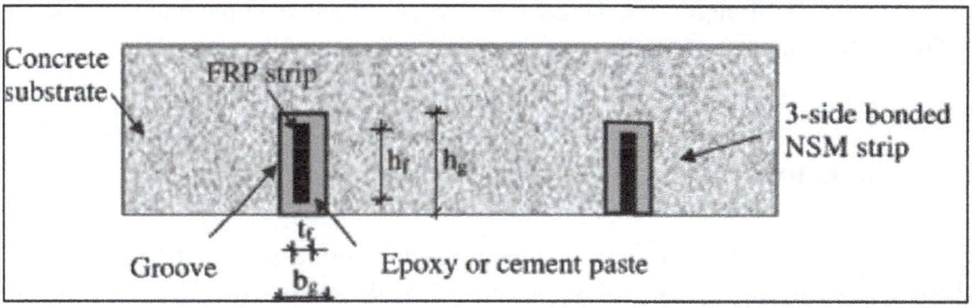

Fig. 8 Schematic section of the MSM system (*Source* Lorenzis and Teng 2007)

Karzad et al. (2017) found that the external bonding of carbon fibers to resist shear stresses was effective both in beams without stirrups and with a minimum amount of stirrups.

Several studies of reinforced concrete beams being subjected to the bending test, and later reinforced with carbon fibers with different types of resin, resulted in significant load increases (Souza 2017). Schneider et al. (2017) tested beams subjected to bending with straight anchoring and "U" anchoring at the end and found that when the straight anchoring has a defined length according to the ACI 440R, the reinforcement with the "U" anchoring does not add rigidity or resistance. According to Deng et al. (2013) the use of the flexural reinforcement in conjunction with the shear reinforcement leads to a better performance of the structural system when compared to the flexural reinforcement alone, due to an increase in stiffness and a better crack control.

The use of the NSM technique reduces the possibility of delamination occurring (Lorenzis and Teng 2007). The adhesion study, based on pullout tests, carried out by Cruz and Barros (2004) already indicated a greater adherence of the NSM application system in relation to the EBR technique. When comparing the EBR and NSM systems, it was observed that the load for the beginning of cracking and the rupture load are greater in the NSM system when using the same area of carbon fiber, indicating a better use of the reinforcement when the blades are inserted in the coverage section (Leoni 2016; Khalifa 2016; Zhang et al. 2017).

As for the use in columns, the confinement effects have different behavior in short and slender columns. Santos et al. (2013) studied reduced models of short columns reinforced by confinement with 1, 2 and 3 layers of carbon fiber blanket and observed that the increase in the load capacity is not linear and that the rupture has always occurred in the seam of two blankets. The authors also compared the laboratory results with some prediction models, concluding that the analyzed models overestimate resistance. The slender pillars were studied by Mendes et al. (2017) working with reduced models with 1, 2 and 3 layers of reinforcement, where they observed that the increase, in the number of layers, increases the rigidity of the structural element but does not increase the capacity of charge. Silva et al. (2019) present a parametric study for dimensioning columns according to the ACI 440R standard, where they observed that the strength of concrete has a great influence on the number of layers of the reinforcement.

7 Conclusions

This work sought to present some techniques and materials used in the rehabilitation of constructions, leaving the professional to choose the method that best suits his case. The techniques of repair systems in concreting failures require the correct choice of material to be used so that there are no micro cracks in the material interface, a fact that would harm the repair system. Injection techniques, on the other hand, must be applied with appropriate equipment and qualified personnel for their correct performance. Surface protection techniques are easier to apply, but they are not the solution for all problems. For the protection against corrosion of the armatures from a carbonation or chloride penetration process, electrochemical techniques are presented as an efficient solution. And for the reinforcement of structures, the polymer composites reinforced with carbon fibers are presented as an alternative of fast and effective execution and that presents durability in face of the environmental action.

References

Abu-Tair AI, Lavery D, Nadjai A, Rigden SR, Ahmed TMA (2000) A new method for evaluating the surface roughness of concrete cut for repair or strengthening. Constr Build Mater 14:171–176

Almeida E et al (2006) Anticorrosive painting for a wide spectrum of marine atmospheres: Environmental-friendly versus traditional paint systems. Progress Organic Coatings 57:11–22

American Concrete Institute (2015) ACI Structural Journal ACI. Guide for the Design and Construction of Structural Concrete Reinforced with Fiber- Reinforced Polymer (FRP) Bars. New York. ACI. 4401R-15

American Concrete Institute (1994) Guide to shotcrete. Detroit, ACI 506 R-90

Baltazar L. et al. Surface skin protection of concrete with silicate-based impregnations: influence of the substrate roughness and moisture. Construction Building Mater 70:191–200

Bethencourt M, Botana FJ, Cano MJ, Osuna RM, Marcos M (2003) Degradation mechanism of an acrylic water-based paint applied to steels. Progress Organic Coatings 47, ed 2, pp 164–168

Callister Jr, WD, Rethwisch DG (2012) Ciência e Engenharia dos Materiais: Uma Introdução. LTC, v. 8° Edição

Chess PM, Broomfield JM (2014) History and principles of protection for reinforced concrete. In: Chess PM, Broomfield JP (eds) Cathodic protection of steel in concrete and masonry. CRC Press, Boca Raton FL

Cruz JS, Barros J (2004) Modeling of bond between near-surface mounted CFRP laminate strips and concrete. Comput Struct 82(17–19):1513–1521 (2004). https://doi.org/10.1016/j.compstruc.2004.03.047

Curcio F, Deangelis B, Pagliolico S (1998) Metakaolin as a pozzolanic microfiller for high-performance mortars. Cem Concr Res 28(6):803–809

Cusson D, Mailvaganam N (1996) Durability of repair materials. Concrete Int-Design Constr 18(3):34–38

Davies K, Broomfield JP (2014) Cathodic protection mechanism and a review of criteria. In: Chess PM, Broomfield JP (eds) Cathodic protection of steel in concrete and masonry. CRC Press, Boca Raton FL

Deng H, Zhang R, Valenca R, Jin J, Fu Q, Bilotti E, Peijis T (2013) Strain sensing behavior of elastomeric composite films containing carbon nanotubes under cyclic loading. Composites Sci Technol 74(24):1–5

Dyer T (2014) Concrete durability. CRC Press, Boca Raton FL

Elbakry HMF, Tarabia AM (2016) Factors affecting bond strength of RC column jackets. Alexandria Eng J 55(1):57–67

Feng-Quing et al (2011) Preparation and properties of an environment friendly polymer-modified waterproof mortar Construction and Building Materials 25:2635–2638

Gomes MA (2015) Propriedades Mecânicas de Compósitos Poliméricos Reforçados com Fibras de Folhas de Abacaxizeiro (Palf). Tese de Doutorado. UNIVERSIDADE ESTADUAL DO NORTE FLUMINENSE DARCY RIBEIRO—UENF CAMPOS DOS GOYTACAZES—RJ AGOSTO

Goyal A et al (2018) A review of corrosion and protection of steel in concrete. Arabian J Sci Eng. Saudi Arabia, pp 5035–5055

Helene PRL (1988) Manual Prático e Reforço de Estruturas de Concreto. São Paulo: Pini

Júlio ENBS, Branco FAB, Silva VD (2004) Concrete-to-concrete bond strength. Influence of the roughness of the substrate surface. Constr Build Mater 18(9):675–681

Karzad AS, Toubat SA, Maalej M, Estephane P (2017) Repair of reinforced concrete beams using carbon fiber reinforced Polymer. MATEC Web of Conferences, volume 120, 01008, 2017. 10.1051

Koumoulos EP, Trompeta AF, Santos RM, Martins M, Santos CMD, Iglesias V, Böhm R, Gong G, Chiminelli A, Verpoest I et al (2019) Research and development in carbon fibers and advanced high-performance composites supply Chain in Europe: a roadmap for challenges and the industrial uptake. J Compos Sci 3:86

Khalifa AM (2016) Flexural performance of RC beams strengthened with near surface mounted CFRP strips. Egypt. Alexandria Eng J 55:pp 1497–1505

Lachovicz PO (2020) Influência do tipo de malha metálica e do tipo de argamassa de realcalinização eletroquímica de estruturas de concreto carbonatado. Dissertação de mestrado, UTFPR, Curitiba, Pr

Lazo EM, Rinchon JPM (2018) Hybrid artificial intelligence-based bond strength model of CFRP-lightweight concrete composite. MATEC Web Conf. Volume 192, 2018 The 4th International Conference on Engineering, Applied Sciences and Technology

Leoni RI (2016) Reforço de estruturas utilizando fibras de carbono: comparação do desempenho de vigas reforçadas com as técnicas EBR e NSM. 2016. 72 f. Monografia (Especialização)—Programa de Educação Continuada da Escola Politécnica da Universidade de São Paulo, São Paulo

Lorenzis L, Teng JG (2007) Near-surface mounted FRP reinforcement: An emerging technique for strengthening structures Hong Kong. Composites: Part B 38:119–143

Macdonald S, Broomfield J (2003) Repairing damaged concrete. In: Macdonald S (ed) Concrete building pathology. Blackwell Science Ltd, Oxford

Mietz IJ (1995) Electrochemical realkalisation for rehabilitation of reinforced concrete structures. Materials Corrosion 46:527–533

Mendes AP, Clivatti N, Araújo SK, Mazer W (2017) Análise da eficiência de compósitos de fibra de carbono utilizados como reforço estrutural em pilares esbeltos de concreto armado. Sodebras 12(139):127–131

Moreno JR, Selmo SM, de S (2007) Aderência de argamassas de reparo de estruturas de concreto. Boletim técnico da escola politécnica da USP. São Paulo

Nanni A (2000) Carbon fibers in civil structures: rehabilitation and new construction. Proc the global outlook for carbon fiber 2000, Intertech, San Antonio, Texas, December 4–6, 6 p

Oliveira G, Alves L (2018) Uso de argamassa polimérica na recuperação estrutural. Boletim de Gerenciamento, 56–64

Parmar R, Rabara D, Rana P, Machhi U, Panchal U, Kansagra M (2018) Use of polymer composite in bridge rehabilitation. 2º International conference on current research trends in engineering and technology, IJSRSET vol 4 Issue 5 Print ISSN: 2395–1990| Online ISSN: 2394-4099

Poletto M (2017) Compósitos termoplásticos com madeira—uma breve revisão. Revista Interdisciplinar de Ciência Aplicada, Petrópolis 4(2):42–482017

Raupach M, Büttner T (2014) Concrete Repair to EN 1504. CRC Press, Boca Raton FL

Santos GMS, Ross HLS, Habitzreuter L (2013) Análise da eficiência do reforço estrutural com fibra de carbono em pilares curtos. Monografia de graduação, UTFPR Curitiba, Pr

Scarfato P, Russo P, Acierno D (2011) Preparation, characterization, and release behavior of nanocomposite microparticles based on polystyrene and different layered silicates. J Applyed Pollimer Sci. https://doi.org/10.1002/app.34783

Schneck U (2011) Electrochemical chloride extraction. In: Concrete repair, A practical guide, Ed. Grantham, M. G

Sherwood P (1966) Protection against atmospheric corrosion of industrial plant, Anti-Corrosion. Methods Mater 13(4):13–14

Schneider FH, Schultz JL, Wierzbicki LR, Mazer W (2017) Análise Do Desempenho da Ancoragem do Reforço de Fibras de Carbono em Vigas Submetidas à Flexão. Curitiba. Revista Técnico Científica Do Crea-PR 1:1–17

SIKA, Soluções para vedação de infiltrações usando os sistemas de injeção SIKA em estruturas de concreto armado, alvenaria e pedra natural, Catálogo técnico, São Paulo (2020)

Silva MTQS, Tokarski RB, Brunhara CA, Corrêa CJ, Mazer W (2019) Parâmetros de cálculo para pilares medianamente esbeltos com diferentes fck's e camadas de fibra de carbono. Brazilian J Develop. https://doi.org/10.34117/bjdv5n12-254

Souza V, Ripper T (1998) Patologia, Recuperação e Reforço de Estrutura de Concreto. Pini, São Paulo

Souza JVC (2017) Análise da influência de 3 tipos de resinas no reforço de fibras de carbono em vigas de concreto armado rompidas por flexão normal. Monografia de graduação. UNESC

Tadolini SC, Mills PS, Burkhard DR (2018) Tekcrete Fast®: Fiber-reinforced, rapid-setting sprayed concrete for rib and surface control. Int J Mining Sci Technol 28(1):29–34

Techniques Soluções em engenharia, acessado em 05/08/2020. https://techniques.com.br/injecao-de-resinas-epoxi-poliuretano-e-gel-acrilico/

Vasconcelos E, Fernandes S, Aguiar J, Pacheco-Torgal F (2011) Concrete retrofitting using metakaolin geopolymer mortars and CFPR. Constr Build Mater, 3213–3221

Wang W et al (2020) Fabrication of all-dimensional superhydrophobic mortar with enhanced waterproof ability and freeze-thaw resistance. Constr Build Mater 238:117626–117636
Wood JGM, King ES, Leek DS (1990) Concrete repair materials for effective structural applications. Constr Build Mater 4(2):64–67
Yeih W, Chang JJ (2005) A study on the efficiency of electrochemical realkalisation of carbonated concrete. Constr Build Mater 19:516–524
Zhang SS, Yu T, Chen GM (2017) Reinforced concrete beams strengthened in flexure with near-surface mounted (NSM) CFRP strips: current status and research needs China. Composites Part B 131:30–42

Time Variant Assessment of Uncertainties for Various Loads Process During the Service Life of Concrete Structures

Ghanshyam Miyani and Arjun Sil

Abstract Concrete structure may be affected due to antagonistic environment, hence responsible for reducing in their strength, stiffness and reliability. However, in general, this unreliability may damage the safety and design life of the concrete structure, therefore, in order to evaluate the current condition of such structures, time dependent reliability assessment of concrete ageing structure essentially must be considered. In this work, the influence on safety and serviceability of structures alongwith various uncertainties such as time variant loading state [i.e. considering dead, live, seismic and wind loads alongwith their effects in combination] of the structure and the effect of ageing on strength-deterioration phenomenon due to various-environmental aggressive parameters need to be extended and examining that the reliability is vulnerable to the selection of initial component strength, however, strength-deterioration models exhibits less sensitivity to the corresponding component strength within the structural system. The critical component need to be identified before performing, conversely, the system reliability helps to reduce the size of analysis. To keep sustaining of reliability, the effect of periodic inspection and maintenance presented in this study throughout the design life/service period of a structure (in order to examine/show the time variant effects in various selected inspection time intervals that yields valuable results) presented in this study.

Keywords Hazard rate · Deterioration · Loads process · Uncertainty · Aging · Reliability · Series system · Design life · Serviceability

G. Miyani · A. Sil (✉)
Department of Civil Engineering, NIT Silchar, Silchar, Assam 788010, India
e-mail: silarjun@gmail.com

J. M. P. Q. Delgado (ed.), *Durability of Concrete Structures*, Building Pathology and Rehabilitation 16, https://doi.org/10.1007/978-3-030-62825-3_6

1 Introduction

Concrete structure may be affected due to aggressive environmental factors which may finally brings changes in the strength, stiffness and reliability of the structure. These effects may cause of a particular component or system strength to deteriorate over time especially when the concrete is exposed in the aggressive environment due to this accelerate risk of failure of the structure. Deterioration of structure may gradually degrade service of the structure in its functional period and thereby affect the structural safety. These should be considered while evaluating the strength for continued service, a process to provide the quantify evidence of the structure that can withstand in remaining life period of the structure against extreme event with an acceptable level of reliability. There are various classes of buildings and other constructed facilities which gradually degrade in service are either known to have damaged, damaging and thereby affecting on structural safety. Many degrade structures are still in use due to the various social and economic aspect, however, increasing warning on safety issue for such structures. Much research has been conducted on the safety evaluation and damage assessment of existing structures considering aspects such as environmental condition, various types of load and its intensity, quality of periodic maintenance and many other factors that may responsible for degradation of the structure. However, the exact effect of these factors could be difficult to predict. The time-dependent behavior and the associated uncertainties in loading and resistance makes it necessary to evaluate the performance and service life of structures using reliability concepts and methods that essentially reflect the time-dependent characteristics of both the load and resistance. However, Mori and Ellingwood (1993) were among the first investigators to assess time dependent reliability of structures considering both the randomness of resistance and the stochastic nature of load in a systematic manner and used for prediction of the remaining life period of deteriorating concrete structures in nuclear plants, bridges and other civil infrastructure.

However, Mori and Ellingwood was not considered the wind load in account of the reliability analysis of the aging structure. Moreover, in a study of wind loads, Knutson et al. (2010) estimated that in the twenty-first century, hurricane wind speeds may increase by as much as 20% around the world; furthermore, the U.S. Climate Change Science Program (CCSP) found that the frequency of hurricanes may increase as well as hurricane wind speeds. Therefore, a proper methodology is needed for the wind load account in evaluation of the reliability of the structure.

This work suggest a method that could be used to evaluate the time-dependent reliability of aging structures in presence of stationarity in various structural loadings experiencing in aggressive environment such as live load, earthquake load and wind load along with variability in structural deterioration, however, methods used to estimate time dependent reliability of a single component and series structural system considered along with the factors that appear to be evaluated or predicted for remaining extended life period of the structure. In addition, effects of periodic inspection and repair in service life of structure to maintain the desired level of reliability has been estimated and presented.

2 Models for Loads and Resistance

Significant occurrence of load events randomly in time alongwith randomness in their intensity is to be considered while examining the service life or evaluating existing residual strength of the structures. If slowly or steadily varies the load intensity during the interval in which it occurs, its effect on the structure is essentially static and the dynamic reaction of the structure is insignificant. For the purpose of reliability analysis, the load intensity may be treated as constant during a load event. In comparison with dead load, other load events duration is generally short, and such events occupy only a small fraction of the total life period of a structure or the component. For these acceptance, a time variant various structural load could be modeled as a sequence of randomly occurring pulses with varying intensity, S_1, $S_2, \ldots S_n$ at times $\tau_1, \tau_2, \ldots \tau_n$ with a small duration, s, as shown in Fig. 1. The duration of such load event 'S' is sufficiently small that it could be assumed that no change in structural strength during the load event.

Based on this illustration of various load incident, a Poisson point process provides a simple model of important load randomly occurring in time and the probability that a number of N(T) load events arise within the time interval (0, T]could be demonstrated by:

$$P[N(0,T] = n] = \frac{\left[\left(\int_0^T \lambda dt\right)^n . \exp\left(-\int_0^T \lambda . dt\right)\right]}{n!}; n = 0, 1, 2, 3 \quad (1)$$

where P[] is equal to the probability of the event in the bracket and λ is equal to the mean occurrence rate of the load events, and assumed to be constant over time.

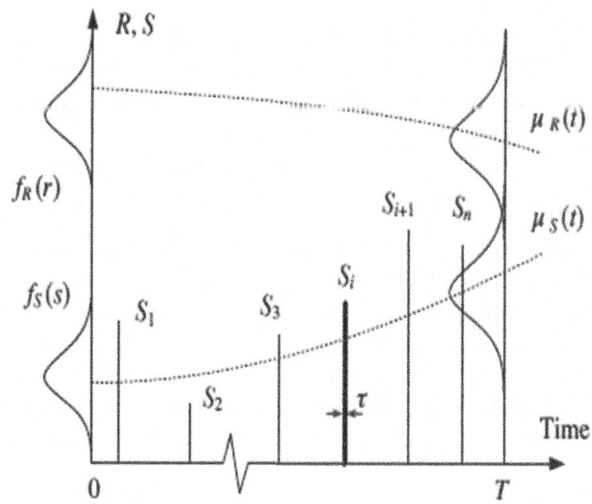

Fig. 1 Schematic representation of load process and degradation of resistance

In the simplest model, the load sequence $S(i, j) = 1, 2, 3, \ldots..n$, is assumed to consist of independent identically distributed (IID) and statistically independent load, with intensities described by the cumulative distribution function (CDF), Fs(x). However, to consider the temporal varying of mean occurrence rate of load, the mean duration τ and the CDF of the load intensity must be treated as time-varying of the structural action arises from the event. The various load statistics are essential to characterize the load process and the load combination used for reliability analysis. However, most building and bridge structures consist of many interconnected individual structural components or elements. In order to consider system reliability, it is important to recognize that the failure of an individual component may or may not represent failure of the whole structure. Alternatively, the reliability of a single member may or may not be representative of the reliability of the entire structural system. The failure probability of the individual element due to insufficient capacity of the element or the occurrence of load effects that exceeds the capacity of the considered element may fail partially or fully, however, could be detected and improved/recovered through periodic inspection and maintenance. Therefore, it is imperative to distinguish between the reliability of individual component and the reliability of the entire system that represents a parallel mode of failure. Hence, evaluation and assessment of failure probability for such structural systems studied by researchers (Enright and Frangopol 1998; Andrzej et al. 2000; Hong 2000) considering system reliability.

3 Time—Dependent Reliability Analysis

3.1 Single Component

For structural component subjected to the sequence of discrete stochastics load events (Fig. 1). However, assuming that structural resistance or strength degrading with time, hence the time-dependent degraded resistance can be expressed as:

$$\mathrm{R(t)} = \mathrm{R_0} * \mathrm{G(t)} \tag{2}$$

where, R(t) is the resistance at time t; R_0 is the initial resistance (a random variable) and G(t) is the degradation function (a stochastic process). The stochastic process model of deterioration described by G(t) is independent of the load history. Models of structural deterioration that have been used in time-dependent reliability analysis of ageing infrastructures are, for the most part, rudimentary.

However, structural deterioration often has been modeled for time-dependent reliability assessment using simple equation in the following form as:

$$\mathrm{G}(t_i) = \mathrm{a} * (\mathrm{T} - t_i)^b \tag{3}$$

where ‘a, and ‘b, are constant parameters determined from an analysis of observed/experimental data. A degradation function depends on two parameters, shape factor and scale factor, which could be selected so as to model to describe complete realistic degradation behaviors. A stochastic degradation model is proposed and subsequently described in this work. The mean value of G(t) decreases in proportion to t^b depending on the deterioration mechanism of interest, such as corrosion of reinforcement (b = 1), sulfate attack (b = 2) and diffusion-controlled aging (b = 0.5). These representative models for degradation process would be considered in the following illustrative examples.

If n number of events occur within time interval $(0,t_L]$ at deterministic times $t_{i,j} = 1, 2, 3,\ldots, n$, the reliability function defined as the probability that the structure becomes reliable through t_L is represent as follows.

$$L(t_L) = P(r.g(t_1) > S_1 \dot{\cap} \ldots\ldots \dot{\cap} r.g(t_n) > S_n] \quad (4)$$

$$L(t_L) = \prod_{i=1}^{n} [Fs(r.g(tj))] \quad (5)$$

$$L(t) = \int_0^{\infty} \exp(-\lambda s.[t - \int_0^{t} Fs(R_0.g(t))dt]).\, f_{Ro}(r).dr \quad (6)$$

If initial resistance Ro is considered deterministic then reliability becomes

$$L(t) = e^{-} \int_0^{T} \lambda(t)[1 - Fs(Ro^{*}g(t))]dt \quad (7)$$

The limit state probability of the failure is given by

$$F(t_L) = 1 - L(t_L) \quad (8)$$

The hazard function or conditional failure rate, expressed as h(t), and defined as the probability of structural failure during interval (t, t + dt] given that the structure survives through (0, t]. The hazard function could be written as:

$$h(t) = -\frac{d}{dt}\{\ln[L(t)\} \quad (9)$$

where L(t) denotes the structural reliability (probability of survival) up to time t. In the presence of structural deterioration, h(t) characteristically increases with time. Once the hazard function/rate is obtained, the time-dependent structural reliability is estimated as (Wang et al. 2015),

$$L(t) = e^{-\int_0^t h(\tau)d\tau} \quad (10)$$

The hazard function at time t, h(t) can be expressed as

$$h(t) = \frac{P(T_n \leq t + dt|T_n > t)}{dt} = \frac{P(t < T_n \leq t + dt)}{P(T_n > t)dt} \tag{11}$$

In which T_f is a random variable denoting the time at which failure occurs, $P(t < T_f \leq t + dt)$ is the probability that failure occurs at time interval $(t, t + dt]$, due to the structure failure, and $P(T_f > t)$ is the probability that the structural member has reliable up to time t, assuming that the occurrence of load events is described by a Poisson process and that R_0 and $g(t)$ are deterministic, we have

$$P(t < T_n \leq t + dt = \lambda(t)dt.P\left[Ro^*g(t + dt) < Q(t + dt)\right].P[t < T_n]. \tag{12}$$

Substituting Eq. 9 into Eq. 10, the hazard function is written as:

$$h(t) = \lambda(t).[1 - F_S(R_0.g(t)] \tag{13}$$

In which $F_S(\dot{})$ is the CDF of the load intensity of failure due to structure agingh(t), increases with time.

4 Selection of Loads and Load-Combination Procedure

The cumulative-time failure probability of a deteriorating element subjected to two statistically independent load processes with intensities S_1 and S_2, where S_1 is time-variant and S_2 is time-invariant could be expressed as:

$$L_s(t) = \int_0^{\infty}\int_0^{\infty} \exp(-\lambda_{S1}.[t - \int_0^{t} F_{S1}(R_0.\, g(t) - S_2)dt)]).\, f_{S2}(S2).\, f_{Ro}(r)\, ds_2.\, dr. \tag{14}$$

In which λ_{S1} = mean occurrence rate of S_1 and F_{S1} and f_{S2} are the CDF and PDF of S_1 and S_2 respectively. If R_0 and S_2 are considered deterministic then reliability becomes

$$L_s(t) = e^{-\int_0^T \lambda_{S1}[1 - F_{S1}(Ro^*G(t) - S_2)]dt} \tag{15}$$

When more than one time dependent loads acting on the structure, the probability distribution of the structural action due to the combined loads could be obtained approximately as function of the statistical description of individual load (Wen 1977; Hwang et al. 1985). At any instantaneous time structure may be subjected to only one load combination. Thus the probability of failure of the structure during its service life period could be estimated by the sum of failure probabilities of this mutually explosive load events combination.

For example, component subjected to two time varying loads with their intensities S_1 and S_2, then the limit state probabilities under the following three exclusive loads combination must be evaluated S_1, S_2 and $S_1 + S_2$. The failure probability P_f could be expressed as

$$P_{f(q)} = P_{f(S1)} + P_{f(S2)} + P_{f(S1+S2)} \tag{16}$$

Above equation also expressed as,

$$F_{qmax(q)} = P_f\left[(S_1 \le q)\right].\ P_f\left[(S_2 \le q)\right].\ P_f\left[((S_1 + S_2) \le q)\right] \tag{17}$$

Approximate substitution of Eqs. 14 and 15.

$$F_{qmax(q)} = \{\, e^{-\int_0^T \lambda_1 [1 - F_{S1}(q)]dt} \}.\{\, e^{-\int_0^T \lambda_2 [1 - F_{S2}(q)]dt} \}.\{\, e^{-\int_0^T \lambda_{12} [1 - F_{S12}(q)]dt} \} \tag{18}$$

where, $F_{qmax(q)} = P_{f(q)}$ is a limit state probability under the load combination q. (q $= S_1 + S_2$), $= \lambda_{12}\lambda_{S1+S2}$(the occurrence rate of the combined load).

The mean occurrence rate of the load combination of $S_1 + S_2$ approximately evaluated as,

$$\lambda_{S1+S2} \approx \lambda_{S1}.\lambda_{S2}.(\tau_{S1} + \tau_{S2}) \tag{19}$$

The mean duration of the load combination $S_1 + S_2$ approximately evaluated as,

$$\tau_q = \frac{(\tau_{S1} * \tau_{S2})}{(\tau_{S1} + \tau_{S2})} \tag{20}$$

In which λ_{Si} and τ_{Si}=mean occurrence rate and the mean duration of the load event S_i.

5 Structural System

A structural system made of various types of element such as beam, column, slabs and wall. However, mainly two types of the structural system such as system in series and parallel systems. In parallel system all of the members must fail before the system fail, for series system the failure of one member lead to immediately failure of the entire system. In reality, most structure cannot be classified as either series or parallel system. They may not fail when a single member fails, but they can fail before all members fail. So the reliability of a structure modeled as a series system of components provides a conservative lower bounds estimate of the actual system reliability.

Let us assume that the 'n' discrete load events occur at the deterministic times $t_i = 1, \ldots, n$ load S_i acting on the structural system. The S_i are the statistically independent and identically distributed random variable. Due to this load, induces the structural action in member i (e.g. moment, axial force, shear, etc.) C_iS_i. The degradation function of the each system element is described in time by the $g_i(t_i)$. The initial strength of the component $r = \{r_1, r_2, \ldots\ldots\ldots r_n\}$ is deterministic. The time dependent reliability function for the series system is represented as,

$$L_{s(t_L)} = P[(r_1 g_1(t_1) > C_1S_1 \cap r_2g_2(t_2) > C_2S_2) \cap \ldots\ldots \cap r_n g_n(t_n) > C_nS_n] \quad (21)$$

$$L_{s(t_L)} = P(r_1 g_1(t_1) > C_1S_1).\, P(r_2g_2(t_2) > C_2S_2)\, P(r_n g_n(t_n) > C_nS_n). \quad (22)$$

$$L_{s(t_L)} = [1 - P(r_1 g_1(t_1) > C_1S_1)].\,[1 - P(r_2g_2(t_2) > C_2S_2)]\ldots[1 - P(r_n g_n(t_n) > C_nS_n)] \quad (23)$$

$$L_{S(t_L)} = \prod_{i=1}^{n} [e^{-\int_0^T \lambda[1 - F_{s1}(ri.gi(t))]dt}] \quad (24)$$

$$L_{s(t_L)} = [e^{-\int_0^T \lambda[1 - Fs1(ri.gi(t))]dt}]^n \quad (25)$$

The above equation for reliability of the ith component follows series system.

6 Component Reliability

The component reliability function effects due to the component strength degradation that could be represented using the several simple parametric representation of time dependent strength. Here estimating those parameters have significant impact on the component reliability. The probabilistic load models having load intensity used in this parametric sensitivity are given in the Table1.

The assumption for the initial strength has lognormally distributed with mean $\mu_R = 1.15\ R_n$, R_n is the nominal strength. And co-efficient of variance $V_R = 0.15$. These statistical data are selected to be representative of existing strength data for

Table 1 Statistical description of structural load

Type of load	Mean	C.O.V	Pdf	frequency (occurrence rate)	τ
Dead load	1.00 D_n	0.07	Normal	–	40 year
Live load	0.40 L_n	0.50	Type 1	0.5	3 months
Earthquake load	0.08 E_{ss}	0.85	Type 2	0.11	30 s
Wind load	0.78 W_n	0.37	Type 1	0.02	3 s

D_n, L_n and W_n= Nominal load and E_{ss}= Safe shout down earthquake

Table 2 Various degradation model

Shape of the degradation function	Degradation rate g(40)	Corresponding degradation mechanism
Linear: $g(t) = 1 - at$	0.7, 0.8, 0.9	Corrosion
Parabolic: $g(t) = 1 - at^2$	0.8, 0.9	Sulfate attack
Square root: $g(t) = 1 - at^{0.5}$	0.8, 0.9	Diffusion-controlled degradation

the reinforced concrete component subjected to flexure or compression. In analysis reference, life period of the structure t_L is taken 60 years, the initial life period of 40 years and an extra 20 year life period for continued service of the structure. Dead load is treated as deterministic because of variability in dead load found in analysis is negligible compared to the other load.

The various degradation models given in Table 2 examined, however, the effect on the time dependent reliability due to the various type of degradation. The resistance degradation is reflected through the rate parameter derived from the ratio of the residual resistance at the end of reference time to the initial resistance, assumed to be 0.7, 0.8 and 0.9 at 40 years. The linear function represents the corrosion effect on reinforcement. The parabolic and square root degradation models assume for sulfate attack and diffusion control respectively.

The combination of nominal live load L_n and dead load D_n is considered according to the following design equation (ACI, Building code 1989).

$$0.9R_n = 1.4D_n + 1.7L_n \tag{26}$$

where, R_n= nominal strength D_n = nominal dead load and L_n = nominal live load for simplicity, it is assumed that $D_n = L_n$. The structure is subjected to both live load and dead load, and since the dead load is assumed to be deterministic, whose effect can be subtracted from the resistance.

The combination of nominal live load L, earthquake load and dead load D is evaluate according to the following design equation (ACI, code 1976).

$$0.9R_n = D_n + L_n + E_{ss} \tag{27}$$

The combination of nominal live load L, dead load D and wind load W is evaluate according to the following design equation (ASCE 7–95).

$$0.9R_n = 1.2D_n + 0.5L_n + 1.3W_n \tag{28}$$

In which E_{ss}= safe-shutdown earthquake loads; W_n=nominal wind load. With structural actions due to E_{ss} and W_n equal to 2 D_n and 3 D_n respectively, the central factors of safety (based on the mean values) are 3.33 for the D + L + E and 4.6 for the D + L + W. The mean occurrence rate for the combination L + E is $\lambda_{L+E} = 0.014$ and the mean occurrence rate for the combination L + W is $\lambda_{L+w} = 0.0025$ calculated as per (Eq. 19).

The effect of the degradation in 40 years on the failure probability is present in (Fig. 2), for D + L, D + L + E and D + L + W utilizing the linear shape of degradation model.

For each degradation rate,F(t), increases nearly linearly with time after 20 years. Above (Fig. 2) shows that failure probability increases with the increase rate of degradation function. Based on the above figures shows that the probability of failure becomes more with all combinations such as D + L, D + L + E and D + L + W load combinations.

However the failure probability under the D + L + E has less susceptibility to the degradation rate and the shape of the degradation function. The failure probability under D + L + E combination with the linear degradation model having g(40) = 0.7, increases by a factor of five and above to that without degradation. The probability of failure under D + L increases by more than two orders of magnitude and above that

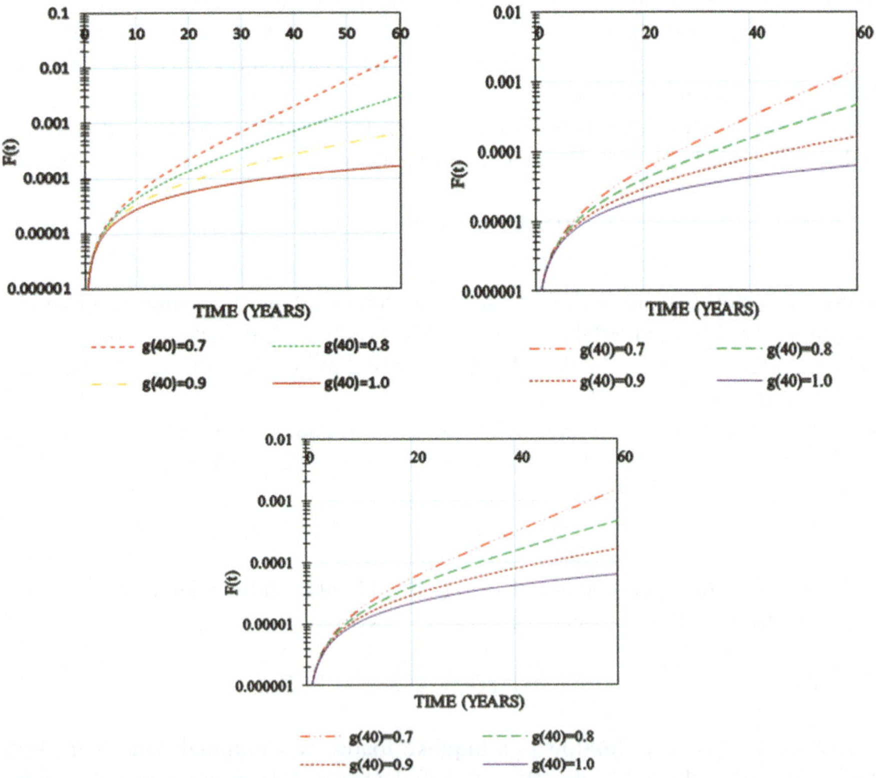

Fig. 2 Dependence of single-component failure probability on degradation rate for various load combinations, **a** D + L; **b** D + L + E; **c** D + L + W

the non-degradation rate. So, the effect of the shape of the degradation (D + L + E) is neglected. This less-sensitivity of the probability of failure under D + L + E to the degradation model is due to the small mean occurrence rate ($0.08E_{ss}$) and large variability/uncertainty in the earthquake load (0.85) process while occurring earthquake load event, therefore, it is highly uncorrelated event/process to identify/evaluate the degradation rate with respect to the normal degradation mechanism of the component whether the component has degraded or not. The probability of failure under the D + L + W is less to the D + L due to the small mean occurrence rate of the wind load. Therefore, D + L is one the most appropriate combinations among all others for estimating the time variant reliability of aging structures.

In this study (from Fig. 2), It has been observed that the probability of failure under load combinations of D + L + E, D + L + W shows less sensitive with the degradation models comparing to the probability of failure under D + L load combination, which shows higher probability of failure or higher aging effect. This is because due to the rate of occurrences under various load combinations exhibits lesser aging effects compare to D + L load combination.

The effect of shape of degradation function for various load combinations D + L, D + L + E and D + L + W on the limit state probability F(t) is shown in (from Fig. 3), for g(t) = 0.8.The failure probability associated with square root model is highest up to 40 years, followed by F(t) for linear and parabolic models. However the failure probability associated with parabolic model increase rapidly after the 40 years.

The same degradation rate g(40) of residual strength at 40 years the degradation in component strength using the square root model occur in early stage of the reference period the component strength using the parabolic model degrades more rapidly after the reference period. The shape of degradation functions has a greater influenced in g(40) decreases. For the lower residual strength at time t, the failure probability sensitivity is also lower.

The effect of shape of degradation function for various load combinations D + L, D + L + E and D + L + W on the limit state probability F(t) is shown in (from Fig. 4), for g(t) = 0.9.

In Fig. 5a shows that probability of failure associated with various mean occurrence rate of load and rate of degradation rate g(40) = 0.8 followed linear shape of the degradation rate. Up to 20 years of life period of the structure has no influenced by occurrence rate of load on failure probability. After 20 years more failure probability failure associated with more occurrence of rate of the load.

The failure probability associated with non-degrading strength for various mean occurrence rate of the structure is presented in the Fig. 5b and clearly shows that probability of failure increases with the increasing in mean occurrence rate of load.

The hazard functions associated with D + L, D + L + E and D + L + W for various load combinations with different increasing degradation rate (Fig. 6) for g(t) = 0.8. Square root degradation model having highest instantaneous probability failure of single critical component up to 40 years of the life period of the structure after that parabolic degradation model increasing rapidly hazardous of the structure.

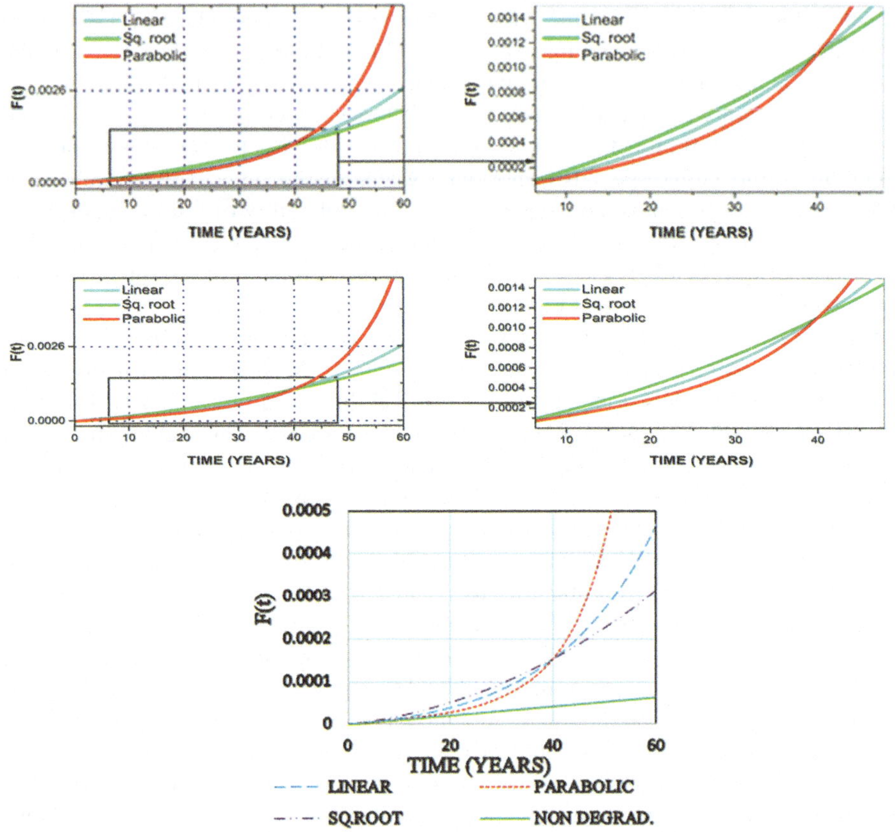

Fig. 3 Dependence of single-component failure probability on degradation model g(40) = 0.8 for various load combinations, **a** D + L; **b** D + L + E; **c** D + L + W

The hazard functions association with D + L, D + L + E and D + L + W for various load combinations with different increasing degradation rate (Fig. 7) for g(t) = 0.9. For resistance degrades linear, parabolic and square root degradation model. The hazardous function associated with the square root strength degradation model is nearly linear. Hazardous function associate with the parabolic and linear strength degradation model are non-linear. Hazard associated with the D + L + W load combination is less than the D + L + E and D + L load combination due to the low occurrence rate of the load event.

The effect of the initiation time t_i is shown (in Fig. 8), for D + L, D + L + E and D + L + W and linear strength degradation expressed as

$$g(t;t_i) = 1; \quad t < t_i \tag{29}$$

$$g(t;t_i) = 1 - a(t - t_i) \quad t \geq t_i \tag{30}$$

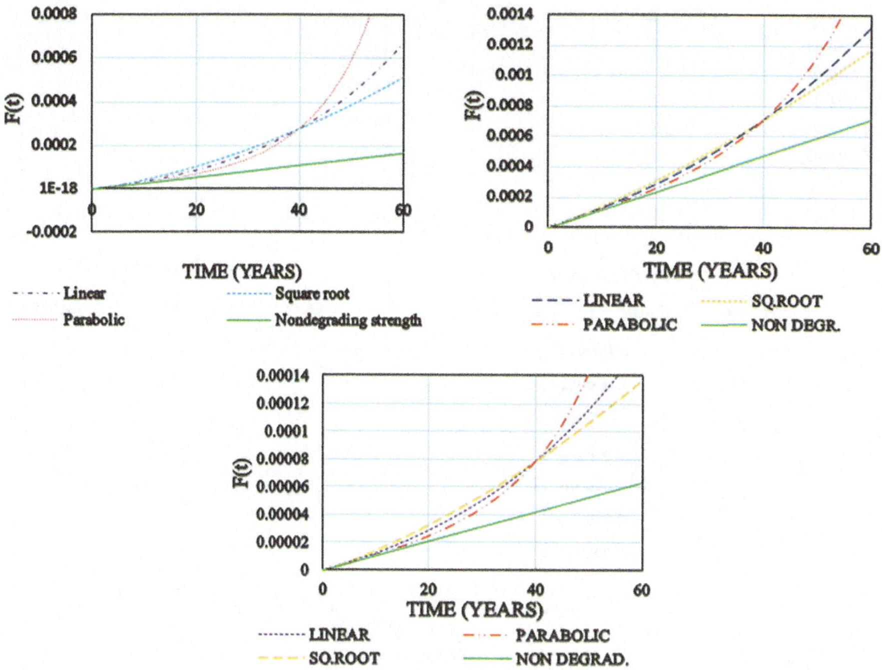

Fig. 4 Dependence of single-component failure probability on degradation model g(40) = 0.9 for various load combinations, **a** D + L; **b** D + L + E; **c** D + L + W

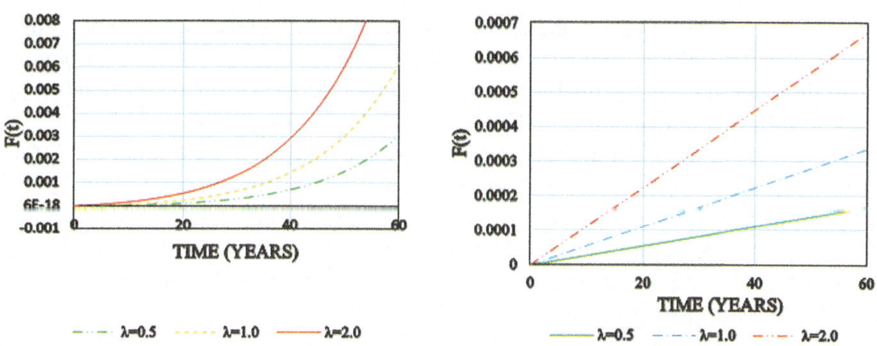

Fig. 5 Dependence of single-component failure probability on mean occurrence rate of load events, D + L: **a** Linear strength degradation; g(40) = 0.8; **b** Non-degrading strength

In the degradation rate is a given by g(40; ti = 0) = 0.8. The failure probability calculated using the linear strength degradation model is increasing mainly in later stage of exposure. Since that is included during the service life depends on the initiation time, the approximate evaluation of initiation time is more important to the reliability assessment in above (from Fig. 8), shows that the initiation time increasing failure probability of single critical component becomes less.

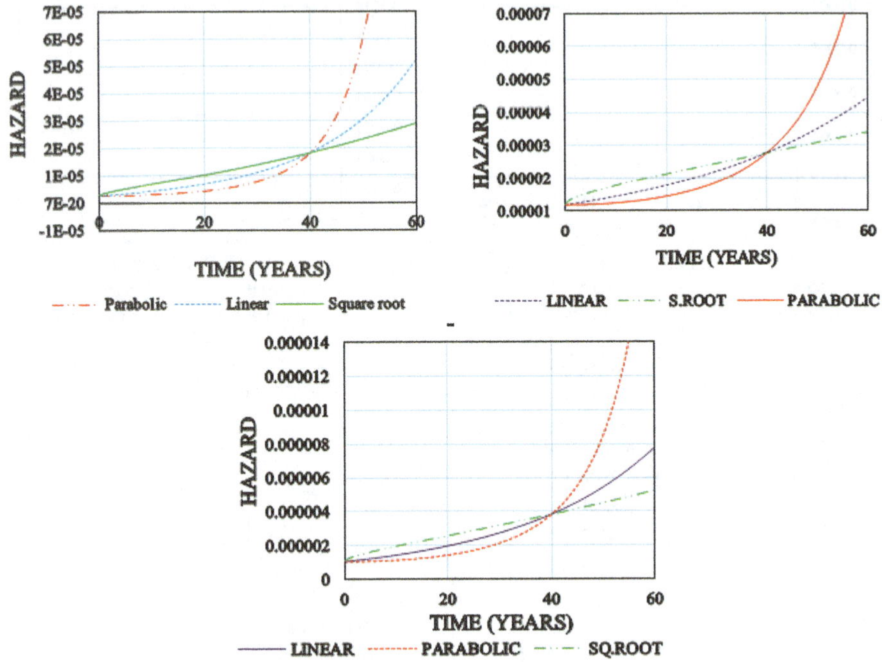

Fig. 6 Hazard function of single component on degradation model g(40) = 0.8 for various load combinations, **a** D + L; **b** D + L + E; **c** D + L + W

A system consist of the components with different initial mean strength of component on system failure as shows in (Fig. 9), and because of this some components affect the system failure probability more than others. Such critical components are a result of basically two sources, difference in initial strength.

The effects of different initial strength of components on system reliability of the structure are shown. The initial strength of all components is increased by 5, 10, 20% and no change in initial strength. Critical component on initial strength remain same. Strength degradation is linearly and follow the g(40) = 0.8. Initial strength of non-critical component is increased 20% the increasing of these component to system reliability is negligible.

In Fig. 10 shows the effect of component degradation rate on system reliability. The degradation rate are g(40) = 0.8 for critical component while g(40) = 0.9, 0.95 and 0.98 for the other non-critical component. Initial strength of all components is statistically independent and lognormally distributed. At the early life period of structure remains same failure probability of structural system except the single critical component. At life period of 40–60 years, the non-critical system reliability and single critical component reliability are approximately same, so it is important to identify the single critical component in the complex system for enable the analysis to keep manageable for the system model.

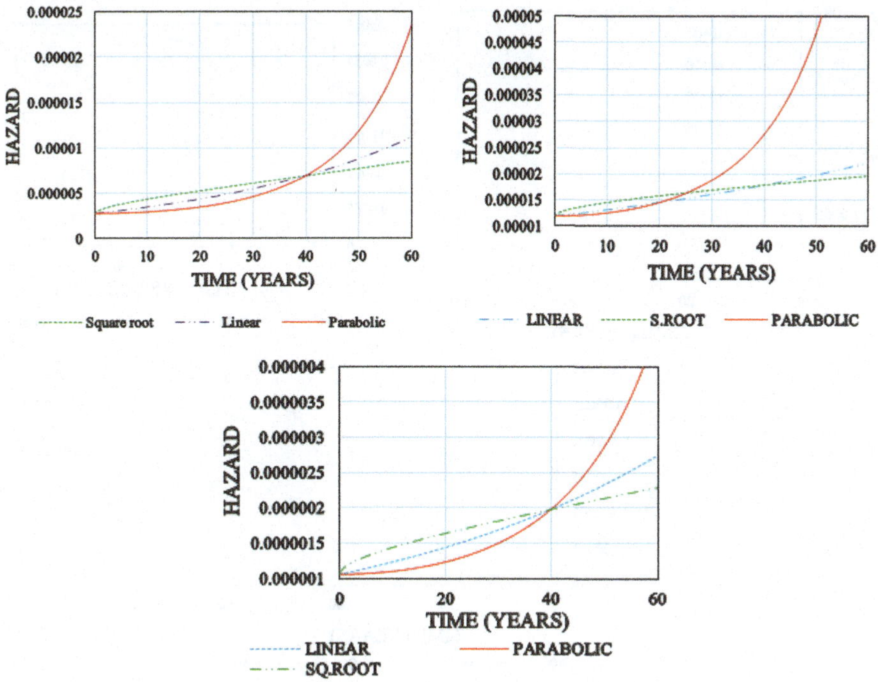

Fig. 7 Hazard Function of single component on degradation model g(40) = 0.9 for various load combinations, **a** D + L; **b** D + L + E; **c** D + L + W

Periodic inspection is suitable to maintenance of the structure for restoring a degraded strength of structure to near original condition. Inspection and maintenance is costly to make it optimum. Prepare strategy for maintenance and inspection periodically of the structure.

In Fig. 11 shows the failure probability with or without inspection for D + L occurrence rate of 0.5/year with strength degradation linearly g(40) = 0.8. However, it has been observed that the time interval of inspection plays a vital role while estimating its current residual strength/capacity of component that may be under or over estimated. In Fig. 11, it can be seen that at the time of inspection and maintenance, the failure probability decreases in slope and becomes more distinct when inspection conducted in different intervals specially lesser interval follows higher reliable that effects in its remaining life period of the component compared to larger interval of inspection, however, cost involvement may be higher in less interval of inspection and hence appropriate inspection interval is thoroughly estimated considering the safety and economy aspects. In strategy 1 component inspection at every 20 year of the life period of the structure and strategy 2 component inspection at every 10 year of the life period of the structure. To make a decision the relation of cost to the extent inspection and maintenance strategy and data of expected losses due the failure is required.

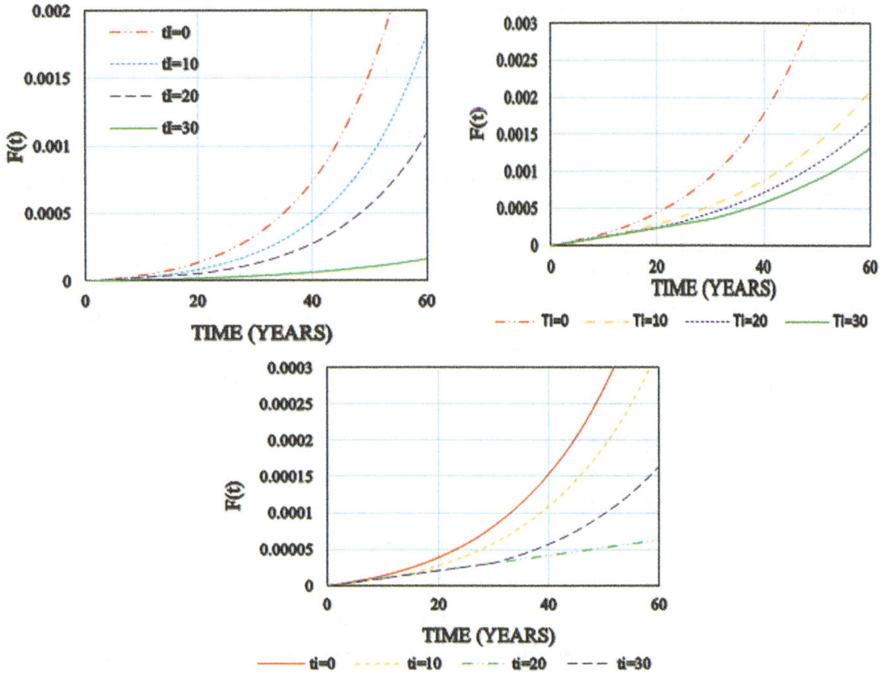

Fig. 8 Dependence of single-component failure probability on initiation time for various load combinations, **a** D + L; **b** D + L + E; **c** D + L + W

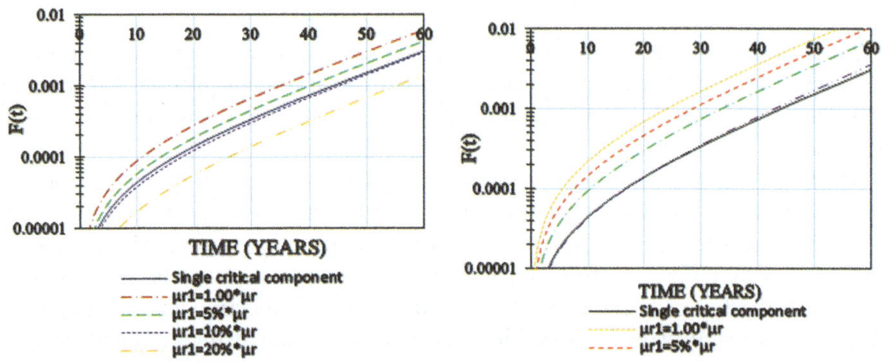

Fig. 9 Dependence of single-component failure probability on initiation strength: **a** Two-component systems; **b** Five-component systems

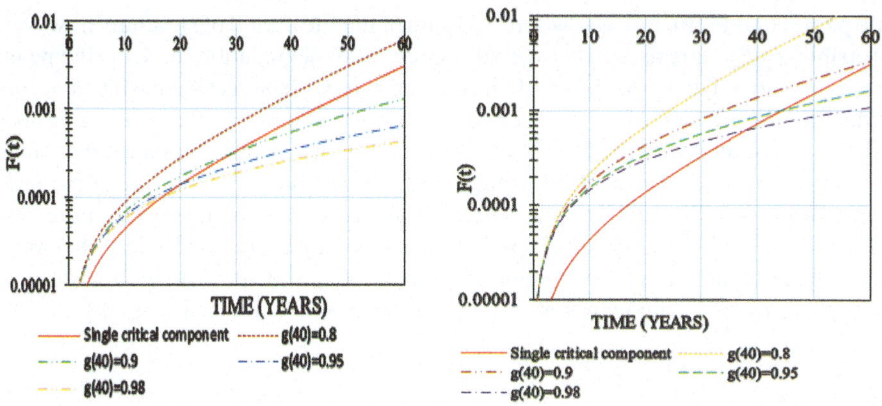

Fig. 10 Dependence of single-component failure probability on degradation rate: **a** Two-component systems; **b** Five-component systems

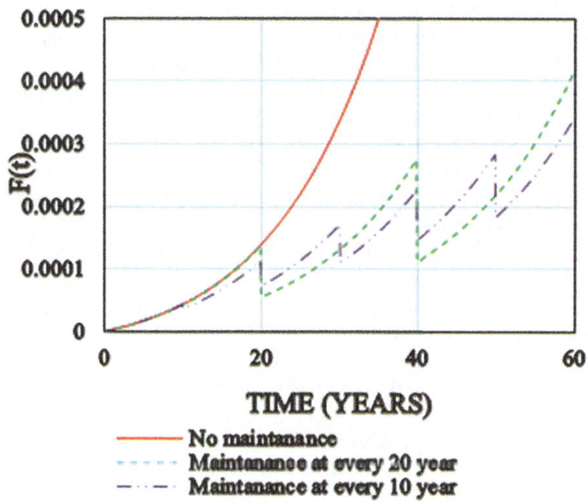

Fig. 11 Failure probability with repair and maintenance

7 Conclusions

This work presented an improved method for the assessment time dependent reliability of component and system. In this study, the sensitivity of component reliability to the various parameters, illustrates load occurrence and strength degradation described by simple examples. The reliability is the susceptibility to the choice of the mean occurrence rate of the load event, component strength and various strength degradation models, including the initiation time and coefficient of variation of time varying load. It was found that probability of failure associated with parabolic degradation model becomes highest among the degradation models after the 40 years of the life period of the structure and within 40 years of the life period of structure, however, probability of failure associated with the square root degradation model is the peak. It was also found that probability of failure increases with increasing the mean occurrence rate of the load. Among three loads combination, the highest failure probability is associated with the D + L. Identification of the critical component becomes important to reduce the reliability analysis of the structural system. Use of the time dependent reliability analysis is to determine the repair and maintenance strategies. In the life period of the structure, periodic inspection provides additional statistical data of the structural condition. For continues service of the structural system or strengthening the component for developing repair and maintenance strategies.

References

Andrzej SN, Kevin RC (2000) Reliability of structures. McGrew Hill USA

Building code requirements for reinforced concrete. (1989) ACI standard 318–89, American concrete Institute (ACI). Detroit, Mich

Code requirements for nuclear safety related concrete structures (1976) ACI standard 349–76, American concrete Institute (ACI). Detroit, Mich

Enright MP, Frangopol DM (1998) Service-life prediction of deteriorating concrete bridges. J Struct Eng 124(3):309–317

Frangopol DM, Lin KY, Estes AC (1997) Reliability of reinforced concrete girders under corrosion attack. J Struct Eng 123:286–297

Hong HP (2000) Assessment of reliability of aging reinforced concrete structures. J Struct Eng 126:1458–1465

Hwang H, Kangami S, Reich M, Ellingwood B, Shinozuka M, Kao CS (1985) "Probability based load combination criteria for design of concrete containment structures", NUREG/CR-3876. Nuclear regulatory commission, Washington, DC, Mar, p 1985

Knutson TR, McBride JL, Chan J, Emanuel K, Holland G, Landsea C et al (2010) Tropical cyclones and climate change. Nat Geosci 3:157–163

Mori Y, Ellingwood BR (1993) Reliability based service life assessment of aging concrete structures. J Struct Eng 119:1600–1621

Wang C, Li, Q-W, Zou A-M, Zhang L (2015) Arealistic resistance deterioration model for time-dependent reliability analysis of aging bridges. J Zhejiang Univ-Sci A (Appl Phys & Eng) 16(7):513–524

Wen Y-K (1977) Statistical combination of extreme loads. Struct Div ASCE 103(5):1029–1093

Fly Ash and Granulated Blast Furnace Slag to Mitigate the Alkali Silica Reaction in Concretes

K. M. V. Moreira, F. R. C. Ribeiro, E. P. de Deus, and A. E. B. Cabral

Abstract The alkali silica reaction (ASR) is a pathological manifestation of chemical origin and high gravity that can negatively affect the mechanical properties and durability of important concrete structures, even with short time in service. The reaction results from the interaction of Portland cement alkalis and other sources (internal and external to concrete) with the reactive silica of the aggregate. Several studies are under development about the best approach for the mitigation of ASR. The use of supplementary cementitious materials (SCM) has been widely accepted. One of the first SCM studied was the fly ash and, in recent years, the granulated blast furnace slag has also been fairly addressed in research. To have the expected efficacy, fly ash and granulated blast furnace slag must present some peculiar characteristics such as high total concentration of silica, alumina and hematite, high fineness and reduced alkalis content. The protection mechanism offered by fly ash and granulated slag involves not only the pozzolanic action, but also the possibility of fixing the alkalis in the pozzolanic C-S-H, reducing the CaO/SiO_2 ratio of the C-S-H formed by the Portland clinker and reducing the permeability of the concrete. Currently, the use of SCM is still the best approach in the ASR mitigation because, in addition

K. M. V. Moreira (✉)
Engineering and Materials Science Postgraduate Program, University Federal do Ceara, Fortaleza, Brazil
e-mail: kelvyamoreira@gmail.com

Civil Engineering Course, University State Vale do Acarau, Sobral, Brazil

F. R. C. Ribeiro
Postgraduate Program of Civil Engineering, University do Vale do Rio dos Sinos, São Leopoldo, Brazil
e-mail: roger.ribeiro_@hotmail.com

E. P. de Deus
Engineering and Materials Science Postgraduate Program, University Federal do Ceara, Fortaleza, Brazil
e-mail: epontes@ufc.br

A. E. B. Cabral
Civil Engineering Postgraduate Program: Structures and Building Construction, University Federal do Ceara, Fortaleza, Brazil
e-mail: eduardo.cabral@ufc.br

J. M. P. Q. Delgado (ed.), *Durability of Concrete Structures*, Building Pathology and Rehabilitation 16, https://doi.org/10.1007/978-3-030-62825-3_7

to offering technical advantages, it also contributes to reducing the environmental impact caused by industries in different segments, either by reducing the CO_2 released into the atmosphere or by reusing by-products.

Keywords Alkali-silica reaction mitigation · Fly ash · Granulated blast furnace slag · Protection mechanism

1 Introduction

The alkali silica reaction (ASR) is one of the expansive chemical reactions of endogenous origin that can occur in concretes throughout its useful life. This reaction results from the combination of alkaline ions Na^+ and K^+ [catalyzed by hydroxyl ions (OH^-)] from Portland cement or other sources with amorphous silica present in certain aggregates under conditions of high humidity.

The reaction's product is an alkaline silicate gel formed in the concrete pore solution. This gel can absorb water and generate hydraulic tensions inside the pores with consequent expansion. As a result, there is the occurrence of cracks and the loss of structural integrity, with a reduction in the properties of mechanical resistance and elasticity modulus.

There are several studies already developed in an attempt to mitigate the deleterious effects of this reaction. Avoiding the use of reactive aggregate and limiting the alkalis content of Portland cement are not technically and economically viable alternatives. Therefore, some researchers have studied the use of lithium-based chemicals and the use of supplementary cementitious materials (SCM) to substitute Portland cement partially during concrete production.

One of the lithium-based chemicals that showed the greatest technical advantages was lithium nitrate ($LiNO_3$). This chemical product guarantees better chemical stability of the reactive silica contained in the pore solution with a lower dissolution rate. However, the complete knowledge of the product mechanisms is still uncertain by researchers, especially in the occurrence of long-term secondary reactions.

The fact is that many researchers defend the use of SCM as the best alternative because they are materials already known chemically and due to the great availability because many are by-products of other industries that need reuse, even though there are also SCM from natural sources and household waste.

The insertion of these SCM in civil construction guarantees economic incentives, such as the reduction in concrete cost due to the replacement of substantial part of the cement with a cheaper material. It also guarantees environmental benefits because the lower demand for cement implies less greenhouse gases released and guarantees a lower amount of industrial by-product and waste disposed in the environment. Skibsted and Snellings (2019) report reductions of 30–40% in CO_2 emissions when replacing Portland clinker with SCM partially without significant change in the material strength and durability performance.

The designation of supplementary cementitious materials is the result of the pozzolanic or cementing activity they exert in the matrix, which may contribute to the Portland cement hydration reactions. Some examples of SCM already consolidated in civil construction are fly ash, active silica, silica fume, metakaolin, rice husk ash and granulated blast furnace slag (American Concrete Institute 2000).

The efficiency of SCM in ASR mitigating depends on specific properties such as fineness, chemical and mineralogical composition and alkalis content. Under this view, fly ash and granulated blast furnace slag are very promising materials.

The following topics discuss in more detail the properties of SCM that can contribute to ASR mitigating, the details of the protection mechanism they offer to concrete and the particular properties of fly ash and granulated blast furnace slag reported by several researchers as a justification for adoption of these materials in ASR mitigating.

2 Supplementary Cementitious Materials Properties

Mostly, the preventive activity of supplementary cementitious materials is simultaneous with its pozzolanic activity, considering that a large number of these materials are pozzolans,[1] such as fly ash, silica fume, metakaolin, ground brick and rice husk ash. Although granulated blast furnace slag has shown positive properties in the ASR mitigation, it is a cementing material that needs an activating alkali (portlandite and gypsum) to form the C-S-H hydration product (hydrated calcium silicate).

Chen et al. (1993) claim that the desirable properties for these materials in order to reduce ASR expansion are: (1) high total concentration of $SiO_2 + Al_2O_3 + Fe_2O_3$; (2) particle size less than 45 μm; and (3) low alkaline content.

The need for a high total concentration of $SiO_2 + Al_2O_3 + Fe_2O_3$ is an inherent characteristic of pozzolanic activity. The higher concentration of these compounds in materials with an amorphous or glassy structure makes the SCM more reactive with portlandite and produces more C-S-H that will fix more alkalis from the concrete.

The fineness of the SCM is essential to ensure a high rate of pozzolanic reaction. The greater fineness of the SCM promotes a greater specific surface. The SCM reaction occurs precisely on the surface of the particles and the fineness favors greater contact with the pore solution (rich in portlandite) and the reaction with the Ca^{2+} ion. Over time and as the reaction progresses, the C-S-H layer formed on the SCM particle surface becomes denser, also favoring the so-called filler effect and contributing to the packaging of the hardened paste microstructure.

It is also important to analyze the alkalis content of the SCM used to ASR mitigation. It will be more favorable to substitute Portland cement with SCM partially

[1] ACI 116R defines pozzolan as a siliceous or siliceous and aluminous material that in itself possesses little or no cementitious value but that will, in finely divided form and in the presence of moisture, chemically react with calcium hydroxide at ordinary temperatures to form compounds having cementitious properties; there are both natural and artificial pozzolans (American Concrete Institute 2000).

that has a lower alkalis content. The American standard that classifies pozzolans, (American Society for Testing Materials 2008), recommends a maximum content of alkalis available in Na_2O of 1.5%.

The fact is that not all alkalis participate in ASR because a portion may remain retained in the crystalline network. Alkalis that can go into solution are the active alkalis. Therefore, it is more assertive to calculate the total active alkalis content of the concrete and establish an acceptable limit value that will vary with the reactivity aggregate.

Some authors have analyzed the contribution of aluminum present in SCM to mitigate ASR. Shafaatian (2012) incorporated aluminum hydroxide [$Al(OH)_3$] into the mixture and effected microstructural studies. The author observed that an aluminosilicate layer with a thickness between 6 μm and 7 μm was created around the aggregate surface preventing the dissolution of reactive silica by hydroxyl ions and resulting in reduced expansion.

Hong and Glasser (1999, 2002) assessed that alumina (Al_2O_3) contributes to fixation of alkalis in C-S-H initially produced and to reduction of alkalinity in concrete pore solution. On the other hand, Chappex and Scrivener (2012a, b) understand that the contribution of alumina is related to the aggregates reactive phases, since the aluminum contained in pore solution inhibits the dissolution of amorphous silica from the reactive aggregates. But Leeman et al. (2015) emphasize that the aluminum present in pore solution only delays the dissolution of the reactive silica and does not promote changes in the morphology, structure and composition of the reaction gel.

3 Protection Mechanism Offered by Supplementary Cementitious Materials

SCM introduced in concretes to ASR mitigation modify: (1) the portlandite content; (2) the Ca/Si ratio of the C-S-H hydration product with increased alkalis fixation; and (3) the hardened paste microstructure, which becomes more compact, with consequences on permeability and mechanical performance. In addition, (4) these materials can have an alkalis dilution effect when used in partial substitution to Portland cement (Olliver et al. 2014).

3.1 Portlandite

The portlandite presence is an important factor influencing expansion due to ASR. The alkaline silicate gel itself is not capable of producing a high expansion pressure. However, the presence of calcium creates conditions that promote the synthesis of a dense gel, such as hydrated calcium silicate (C-S-H). This gel is present in the external zone of the reaction and acts as a semipermeable layer. Water, Na^+/K^+

Fig. 1 Off-white ASR gel in concrete sample by stereoscopic microscopy (12x increase) (Sbrighi Neto et al. 2018)

and OH^- ions can access the reaction site, but the reaction products cannot pass through the Interstitial Transition Zone (ITZ). As a result, there is an increase in pressure between the aggregate susceptible to alkalis and the cement paste. This pressure can reach 20 MPa resulting in micro cracks and progressive degradation of the concrete (Fig. 1). In addition, the alkalis released at the reaction end participate of the deterioration of particles deeper layers of aggregates susceptible to alkalis. This may explain the more severe destructive effect in case of excessive amounts of Na^+/K^+ and Ca^{2+} ions compared to Na^+/K^+ ions only (2014).

Researchers like Kim et al. (2015), Šachlová et al. (2010), Wang and Gillott (1991) agree that the presence of calcium seems to be essential for the ASR expansion. They report the role of calcium in the formation of a stable (non-soluble) variety of ASR gel, suggesting that, in the absence of calcium, silica minerals simply dissolve and remain in the solution. Generally, the calcium content in the ASR gel increases with age and the gels found in the paste have a higher calcium content than those found within the reactive aggregate particles.

SCM react secondarily by consuming portlandite that was produced in the cement initial hydration to form more C-S-H, called pozzolanic C-S-H. Due to the reduction of available $Ca(OH)_2$ there will be less hydroxyl (OH^-) for the reactive silica dissolution. The calcium concentration will form the pozzolanic C-S-H that will fill the matrix pores and ITZ making the alkalis present fix more easily in the C-S-H structure. As a result, the solution alkalinity and aggressiveness will reduce. In addition, the lower Ca/Si ratio due to the consumption of the Ca^{2+} ion will favor the alkalis fixation.

3.2 Ca/Si Ratio

The introduction of SCM in Portland cement paste reduces the Ca/Si ratio of C-S-H causing an important increase in the content of alkalis that these hydrates can fix inside its structure. For example, without SCM, the Ca/Si ratio are generally between 1.5 and 2.0; for pastes aged four years, with 40% Portland cement replaced by fly ash, the value of this ratio decreased to around 1.0. It is possible to say that SCM form pozzolanic C-S-H that extract greater amounts of alkalis from the interstitial solution, decreasing its alkalinity and reducing its aggressiveness towards reactive aggregates (Olliver et al. 2014).

3.3 Alkalis Dilution

An additional measure to inhibit ASR is to reduce the total active alkalis content in the concrete. Clinker is the main source of alkalis in Portland cement and comes from its raw material, although other sources may contribute. Therefore, the substitution of Portland cement by SCM can reduce the total active alkalis content available. Obviously, this measure will only be valid if the SCM alkalis content is lower than the cement alkalis content or if the SCM alkalis are not soluble.

It is important to highlight that the alkaline ions fix more strongly on C-S-H formed by SCM, not being available in the pore solution and helping to reduce the total alkalis content in the concrete.

Metha and Monteiro (2014) remember that the alkalis present in slags and natural pozzolans are insoluble in acid and, probably, will not be available to react with the aggregate. Olliver and Vichot (2014) also state that SCM act as diluents, releasing alkalis more difficultly than cement and making the interstitial solution less aggressive due to reduced alkalinity. However, these authors warn that this effect of alkali dilution made possible by MCS is only applicable in cases where the SCM replaces cement and where the mobilizable alkali content of the SCM is lower than that cement. For them, there is a difficulty in explaining the role of alkali dilution when isolated from the chemical context as a whole, in particular the notion of pessimum content.

3.4 Hardened Paste Microstructure

The use of SCM together with Portland cement also has the advantage of refining the concrete pore solution. The SCM hydration occurs in the pores formed initially during the Portland cement reactions and containing portlandite. The result is the formation of solid products from this hydration and the pore size distribution becomes thinner.

This refinement of the pore structure is particularly important for compounds cements (Portland cement + SCM) because it improves the durability properties of concrete by restricting the growth of hydrates inside the pores that could contain ASR products.

Another positive factor is the fineness of the SCM. The finer material will have a faster hydration reaction speed due to the greater specific surface. This increase in the reaction speed favors the formation of hydrated products more quickly and the guarantee of durability at an early age. The use of SCM together with Portland cement also has the advantage of refining the concrete pore solution. The SCM hydration occurs in the pores formed initially during the Portland cement reactions and containing portlandite. The result is the formation of solid products from this hydration and the pore size distribution becomes thinner.

This refinement of the pore structure is particularly important for compounds cements (Portland cement + SCM) because it improves the durability properties of concrete by restricting the growth of hydrates inside the pores that could contain ASR products.

Another positive factor is the fineness of the SCM. The finer material will have a faster hydration reaction speed due to the greater specific surface. This increase in the reaction speed favors the formation of hydrated products more quickly and the guarantee of durability at an early age.

4 Contribution of Fly Ash

Fly ash is an artificial pozzolan obtained by mechanical or electrostatic means from the combustion gases of furnaces in coal-based thermoelectric plants. The fly ash particles are spherical and have, at least, the same fineness as cement, so that silica is easily available for the reaction (Neville and Brooks 2013).

Fly ash was one of the first SCM exploited in an attempt to mitigate ASR. Saha et al. (2018) suggest that fly ash with low alkali content, high amorphous silica content and low calcium content is the most effective in mitigation. The presence of high content of amorphous silica consumes portlandite from the pore solution causing a reduction in alkalinity. In addition, amorphous silica reduces the concrete porosity by the pozzolanic reaction product resulting in less voids available to accumulate the ASR viscous gel. Shehata and Thomas (2000) corroborate these statements, indicating that, in general, the expansion increases with the increase of the calcium or alkali content of fly ash or with the reduction of the silica content. These authors also identified that the expansions depend not only of the calcium content of fly ash, but also of the Ca/Si ratio. Higher Ca/Si ratios result in greater expansions (Fig. 2).

When analyzing the fly ash effect with the same content of alkalis, but with different calcium content, Shehata and Thomas (2000) identified that the Portland cement substitution percentages capable of reducing expansions in concrete prisms to acceptable limits (0.04% after 2 years) varied. Ashes with low calcium content (6.38% CaO) were very efficient in controlling expansion (Fig. 3a): with 20% substitution, it was possible to reduce the expansion to only 0.043%, indicating that a substitution percentage of 25% could reduce the expansion to less than 0.04%. On the other hand, fly ash with a high calcium content (27.70% CaO) was much less

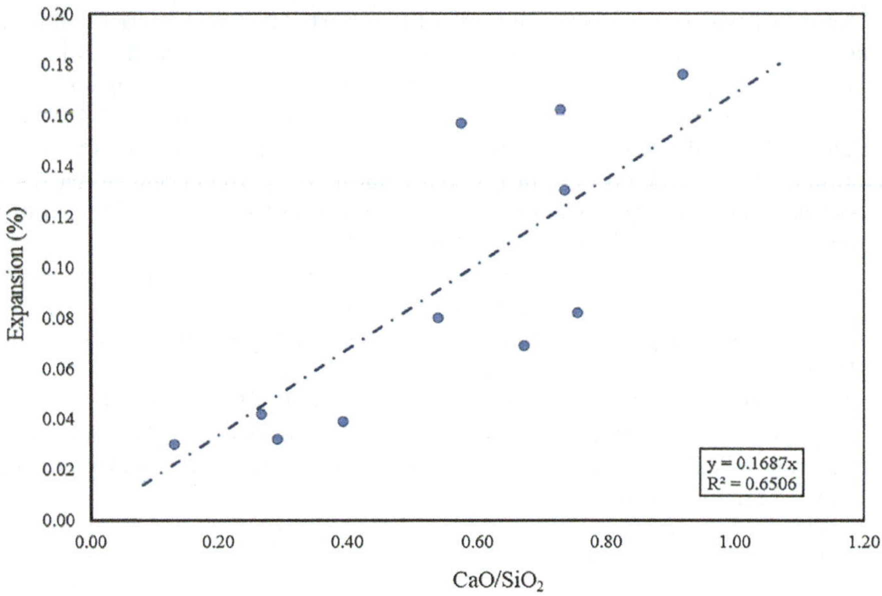

Fig. 2 Correlation between the expansion due to ASR and the Ca/Si ratio of fly ash (Shehata and Thomas 2000)

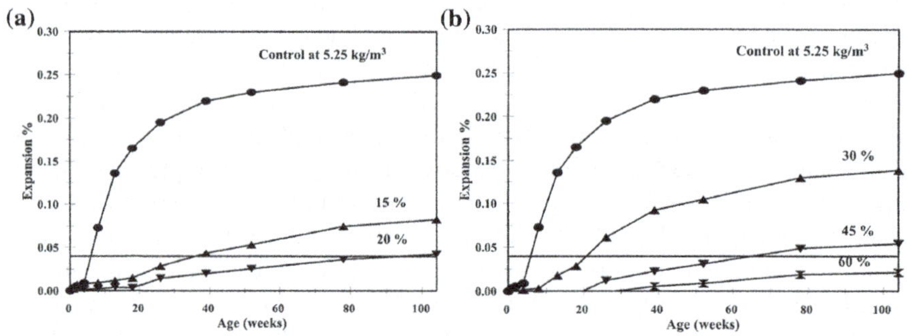

Fig. 3 Fly ash effect with **a** low and **b** high calcium content in expansion due to ASR (Shehata and Thomas 2000)

effective in this role (Fig. 3b), indicating that a substitution percentage between 45% and 60% would be necessary to limit expansion to less than 0.04% with this material.

Kawabata and Yamada (2017) affirm that there is a pessimum content of Portland cement substitution for fly ash that will not guarantee the expansion reduction by ASR due not only to the characteristics of the fly ash (pozzolanic reactivity), but also, of the reactivity of the aggregates in itself. Highly reactive silica minerals, such as opal, cristobalite and tridymite, react with alkalis even in low contents. Aggregates containing these silica minerals often show the pessimum effect (Katayama 1997). Generally, the balance between the alkaline hydroxide content in the pore solution and the reactivity of the aggregates explains this effect.

Kawabata and Yamada (2017) also report that the main contribution of fly ash is to reduce the pore solution alkalinity that depends on the thermodynamic balance between the solution and the C-S-H gel. This C-S-H gel with a low Ca/Si ratio can fix a large amount of alkali. However, when the alkali content exceeds the hydration reaction limit, the C-S-H gel plays a role in supplying the alkali to the pore solution. Thus, the inhibitory role of fly ash in the expansion decreases in a pessimum proportion because the very reactive aggregate can react with alkalis even in low contents.

5 Contribution of Granulated Blast Furnace Slag

Granulated blast furnace slag is a by-product of the manufacturing of iron in a blast furnace where iron ore, limestone and coke heat up to 1500 °C. When these materials melt in the blast furnace, two products are produced—molten iron and molten slag. The molten slag is lighter and floats on the top of the molten iron. The molten slag comprises mostly silicates and alumina from the original iron ore combined with some oxides from the limestone. The process of granulating the slag involves cooling the molten slag through high-pressure water jets. This rapidly quenches the slag and forms granular particles generally no larger than 5 mm in diameter. The rapid cooling prevents the formation of larger crystals and the resulting granular material comprises some 95% non-crystalline calcium-aluminosilicates. The processing of granulated slag is the drying and subsequent grinding into very fine powder, which is granulated blast furnace slag (Siddique 2008).

Therefore, granulated slag is a vitreous material obtained when the molten slag quickly cools and its chemical composition depends, mainly, on the type of iron ore used as raw material. The use of slag in the production of concrete brings advantages already known, such as high mechanical resistance; low permeability; high density; and high resistance to the penetration of chloride ions, the attack of sulfates and corrosion of steel.

The effectiveness of slag in the ASR mitigation has been the subject of several studies in the last decades (Chen et al. 1993, 1994a, b; Thomas 2011; Scott and Alexander 2016). It is clear that, generally, a large volume of cement substitution is necessary to guarantee the ASR mitigation with the use of this SCM. Thomas (2011) reported that more than 50% substitution of cement by slag was necessary to reduce the expansion of concrete prisms below 0.04% after two years of testing with the use of reactive fine aggregate. Similarly, Choi and Choi (2015) found values of 60% substitution of cement to reduce expansion below the permitted limit. Neville and Brooks (2013) affirm that England recommends a common Portland cement composition (Type I ASTM) with a minimum 50% granulated blast furnace slag content.

The researchers present several hypotheses to explain the efficacy of slag in mitigating expansion: effect of alkali dilution, pozzolanic action, trapping alkalis by C-S-H, changes in the Ca/Si ratio of C-S-H, changing the role of calcium hydroxide and decreasing the concrete permeability with cement and slag (Ramachandran 1998). It is noted that, although the slag is classified as a cementing material, and not purely pozzolanic, the mechanisms exposed in the literature for the ASR mitigation are similar to those exposed to fly ash (purely pozzolanic).

The fact is that natural and artificial pozzolans have characteristics of high contents of silica or silica and alumina. Granulated blast furnace slag additionally has a high calcium content, allowing semi-hydraulic reactions to occur even in the absence of Portland cement (Skibsted and Snellings 2019). This high calcium content of the slag may be the negative factor in the purpose of mitigating the ASR, making high contents of Portland cement substitution are necessary to reduce the Ca/Si ratio of the binders and the portlandite content of the concrete after the hydration reactions.

It is necessary to understand that the first step in the study of this SCM with the purpose of ASR mitigation is, therefore, to know its mineralogical composition and crystalline structure because calcium in its free form can be harmful to the concrete, contributing to the ASR, by instead of mitigating (pessimum effect).

As an example, we have the study by Mahyar (2018) who analyzed 11 SCM, including fly ash, granulated blast furnace slag and natural pozzolans, and found that the total weighted contents of CaO and SiO_2 in cement and SCM mixes are strongly related to the expansion caused by ASR in mortars. The increase in CaO and SiO_2, respectively, increases and decreases the ASR expansion in mortars, as well as the increase in Al_2O_3 decreases the expansion, but the relationship is weaker.

However, the viability of granulated blast furnace slag in the ASR mitigation depends on the hematite high concentration (Fe_2O_3) in the vitreous form (obtained by rapid cooling in the production process) combined with the particles fineness, which allows high reactivity with portlandite present in the pores solution of the concrete.

6 Conclusions

The severity of the alkali silica reaction in concrete structures is already consensus among researchers and builders. Buildings with a few years in service present serious structural problems such as reduced mechanical properties and performance.

Several studies have emerged in an attempt to find a way to mitigate this pathological manifestation and the use of SCM has shown good responses, such as fly ash and granulated blast furnace slag.

However, it is important to analyze the use of SCM as an ASR mitigation measure on a case-by-case basis to define the best substitution content for Portland cement by SCM in the manufacture of concrete for a specific set of raw materials. Each reactive aggregate has its own characteristics and particular mineral rates capable of combining with alkalis.

In addition, high substitution content for Portland cement by SCM can result in loss of important mechanical properties of the concrete added to the pessimum effect of some SCM, when the incorporation of more SCM does not result in ASR mitigation.

Although such peculiarities need to be carefully analyzed, the most widespread and promising ASR mitigation measure to date is still the use of SCM. Several researchers have studied other SCM, generally industrial by-products, such as sugarcane bagasse ash, straw ash, rice husk ash.

References

American Concrete Institute (2000) ACI 116R: cement and concrete terminology. Farmington Hills, Michigan

American Society for Testing Materials (2008) ASTM C618: standard specification for coal fly ash and raw or calcined natural pozzolan for use in concrete. ASTM, West Conshohocken

Chappex T, Scrivener K (2012a) Alkali fixation of C-S-H in blended cement pastes and its relation to alkali silica reaction. Cement Concrete Res 42(8):1049–1054

Chappex T, Scrivener K (2012b) The influence of aluminium on the dissolution of amorphous silica and its relation to alkali silica reaction. Cement Concrete Res 42(12):1645–1649

Chen H, Soles JA, Malhotra VM (1993) Investigations of supplementary cementing materials for reducing alkali-aggregate reactions. Cement Concrete Compos 15(1–2):75–84

Choi YC, Choi S (2015) Alkali–silica reactivity of cementitious materials using ferro-nickel slag fine aggregates produced in different cooling conditions. Construct Build Mater 99:279–287

Duchesne J, Bérubé MA (1994a) The effectiveness of supplementary cementing materials in suppressing expansion due to ASR: another look at the reaction mechanisms. Part 1: concrete expansion and portlandite depletion. Cement Concrete Res 24:73–82

Duchesne J, Bérubé MA (1994b) The effectiveness of supplementary cementing materials in suppressing expansion due to ASR: another look at the reaction mechanisms. Part 2: pore solution chemistry. Cement Concrete Res 24(2):221–230

Hong SY, Glasser FP (1999) Alkali binding in cement pastes. Part I—the C-S-H phase. Cement Concrete Res 29(12):1893–1903

Hong SY, Glasser FP (2002) Alkali sorption by C-S-H and C-A-S-H gels. Part II—role of alumina. Cement Concrete Res 32(7):1101–1111

Katayama T (1997) Petrography of alkali-aggregate reactions in concrete—reactive minerals and reaction products. In: East Asia Alkali-aggregate reaction Seminar. Supplementary Papers, pp 45–59

Kawabata Y, Yamada K (2017) The mechanism of limited inhibition by fly ash on expansion due to alkali-silica reaction at the pessimum proportion. Cement Concrete Res 92:1–15

Kim T, Olek J, Jeong H (2015) Alkali–silica reaction: kinetics of chemistry of pore solution and calcium hydroxide content in cementitious system. Cement Concrete Res 71:36–45

Krivenko P et al (2014) Mechanism of preventing the alkali–aggregate reaction in alkali-activated cement concretes. Cement Concrete Compos 45:157–165

Leeman A et al (2015) ASR prevention: effect of aluminum and lithium ions on the reaction products. Cement Concrete Res 76:192–201

Mahyar M, Erdoğan ST, Tokyay M (2018) Extension of the chemical index model for estimating Alkali-Silica reaction mitigation efficiency to slags and natural pozzolans. Construct Build Mater 179:587–597

Mehta PK, Monteiro PJM (Ed. tradução), Hasparyk NP (2014) Concreto: microestrutura, propriedades e materiais. 2.ed. IBRACON, São Paulo

Neville AM, Brooks JJ (2013) Tecnologia do concreto. In: Cremonini RA. (trad.), 2 ed. Brookman, Porto Alegre

Olliver JP, Vichot A, Cascudo O, Carasek H (2014) Durabilidade do concreto: bases científicas para a formulação de concretos duráveis de acordo com o ambiente. IBRACON, São Paulo

Ramachandran VS (1998) Alkali-aggregate expansion inhibiting admixtures. Cement Concrete Compos 20(2–3):149–161

Šachlová Š, Přikryl R, Pertold Z (2010) Alkali–silica reaction products: comparison between samples from concrete structures and laboratory test specimens. Mater Charact 61:1379–1393

Saha AK et al (2018) The ASR mechanism of reactive aggregates in concrete and its mitigation by fly ash: a critical review. Construct Build Mater 171:743–758

Sbrighi Neto C, Quitete EB, Battagin AF (2018) Prática recomendada Ibracon: guia de prevenção da reação álcali-agregado. Ed. Ibracon, São Paulo

Scott A, Alexander MG (2016) Effect of supplementary cementitious materials (binder type) on the pore solution chemistry and the corrosion of steel in alkaline environments. Cement Concrete Res 89:45–55

Shafaatian S (2012) Innovative methods to mitigate alkali-silica reaction in concrete materials containing recycled glass aggregates. Dissertation of MSc. Pennsylvania State University, University Park, Pennsylvania

Shehata MH, Thomas MDA (2000) The effect of fly ash composition on the expansion of concrete due to alkali-silica reaction. Cement Concrete Res 30:1063–1072

Siddique R (2008) Waste materials and by-products in concrete. Springer. e-ISBN: 978-3-540-74294-4

Skibsted J, Snellings R (2019) Reactivity of supplementary cementitious materials (SCMs) in cement blends. Cement Concrete Res 124

Thomas M (2011) The effect of supplementary cementing materials on alkali-silica reaction: a review. Cement Concrete Res 41:1224–1231

Wang H, Gillott JE (1991) Mechanism of alkali–silica reaction and the significance of calcium hydroxide. Cement Concrete Res 21:647–654

Influence of the Addition of Sugar Cane Bagasse Ash on Durability of Concrete Structures

J. M. P. Q. Delgado, R. Berenguer, N. Lima, A. C. Valdés, M. H. F. Medeiros, A. C. Azevedo, and N. B. D. Lima

Abstract The environmental impact of cement production was increased significantly in the last years. It is estimated that for each ton of cement produced, approximately one ton of carbon dioxide is emitted, divided into: decarbonation (50%), combustion in the clinker furnace (40%), transport of raw materials (5%) and electricity (5%). From a world perspective, green strategies have been advanced in order to reduce the environmental impact of the cement production. In this sense, the sugar cane bagasse ash is a by-product resulted from boilers of sugar and alcohol factories. This material, composed mainly of silica, can be used as a mineral admixture in mortar and concrete, with great economic and environmental advantages. This work intent to collaborate for the partial substitution of Portland cement by sugar cane bagasse (SCB), in order to reduce the consume of clinker by volume of concrete, which is the responsible for the high emission of CO_2 to the atmosphere. For this,

J. M. P. Q. Delgado (✉) · A. C. Azevedo
CONSTRUCT-LFC, Departamento de Engenharia Civil, Universidade do Porto, Rua Dr. Roberto Frias, s/n, 4200-465 Porto, Portugal
e-mail: jdelgado@fe.up.pt

A. C. Azevedo
e-mail: antonio.costaazevedo@fe.up.pt

R. Berenguer
Department of Civil Engineering, Catholic University of Pernambuco, 50050-900 Recife, PE, Brazil
e-mail: romildo.berenguer@ufpe.br

N. Lima
Polytechnic School, University of Pernambuco, 50720-001 Recife, PE, Brazil

A. C. Valdés · M. H. F. Medeiros
Department of Civil Construction, Federal University of Paraná, 81531-980 Curitiba, PR, Brazil
e-mail: analietcv@yahoo.es

M. H. F. Medeiros
e-mail: medeiros.ufpr@gmail.com

N. B. D. Lima
Department of Fundamental Chemistry, Federal University of Pernambuco, 50740-540 Recife, PE, Brazil
e-mail: nathalia.blima@ufpe.br

J. M. P. Q. Delgado (ed.), *Durability of Concrete Structures*, Building Pathology and Rehabilitation 16, https://doi.org/10.1007/978-3-030-62825-3_8

an experimental campaign was carried out with cementitious pastes to evaluate the changes in their properties, namely the durability, due to the use of SCB ash. Finally, it could be concluded that concrete structures with sugar cane bagasse ash presented an interesting behavior and, in terms of durability, the samples containing 15% of the ash of sugarcane bagasse showed good results, being important to highlight the need and interest of doing further research on the topic.

Keywords Concrete · Sugar cane bagasse ash · Durability · Chemical analyse · Physical analyse · Sustainability

1 Introduction

The environmental impact of cement production was increased significantly in the last years (Moumin et al. 2020). It is estimated that for each ton of cement produced, approximately one ton of carbon dioxide is emitted, divided into: decarbonation (50%), combustion in the clinker furnace (40%), transport of raw materials (5%) and electricity (5%) (Moumin et al. 2020). From a world perspective, green strategies have been advanced in order to reduce the environmental impact of the cement production. In this sense, researchers reported aspects on the environment impact and sustainable development associated with the cement industry of China (Shen et al. 2017). They concluded that by considering the last decades, the cement industry of China implemented several sustainable strategies in its production steps. For example, solid wastes and recycling materials have been used (Shen et al. 2017). In addition, Hossain et al. evaluated two green strategies to reduce the emission of CO_2 at the cement production by the industries of Hong Kong (Hossain et al. 2017). Their strategies are based on the use of glass powder (obtained from locally generated waste glass bottles) and a bio-fuel prepared from locally generated wood wastes (Hossain et al. 2017). These green strategies reduced 12% of the total of emission gases and 15% of the energy consumption at the cement production (Hossain et al. 2017).

On the other hand, Wilk et al. (2019) reported a solution for the Poland cement industry to estimate the quantity of asbestos–cement products employed at the country. In this article, the authors concluded that the Random Forest algorithm can be applied to this challenge (Wilk et al. 2019). An environmental impact assessment was investigated by Saffari et al. (2019), where the Shahrood cement plant of the Iran was considered. The results revealed that the contaminations due to the plant, especially, were identified for the following systems: the air; the soil of the area; the area of landscape, and the ecology (Saffari et al. 2019).

Aspects of the environmental and economic potential of the use of the cement of limestone-calcined clay in Cuba were investigated (Sánchez Berriel et al. 2016). In this work, the potentiality of the calcined cement was compared with the commercial zeolite cement (Sánchez Berriel et al. 2016). The authors concluded that the limestone calcined clay cement has potential to be commercialized, and a low quantity of CO_2

at its processing steps. In addition, this more sustainable cement can be produced in other countries since the distance from the clay deposit to the cement plant be until 100 km if trucks are as a means of transport.

Recently, advances to improve the more sustainable cement production were reported (Shirkhani et al. 2018; Chen et al. 2010; Miller and Myers 2020; Hossain et al. 2018; Sagastume Gutiérrez et al. 2017; Song et al. 2016; Vázquez-Rowe et al. 2019; Wu et al. 2018; Carvalho et al. 2018). For example, Chen et al. (2010) employed the life cycle impact assessment to investigate aspects of different processes and cement plant variabilities. The results revealed differences between the cement production processes in the set of the studied plants, for example, the authors concluded that the variations are greater than 40%, mainly, due to the lack of accurate measurements on pollutant content (Chen et al. 2010). This result indicates that studies associated with aspects of novel green strategies that are capable to reduce the environmental impact of cement production are important.

The use of the cement in the processing of construction materials, such as the concretes, also can lead to the environmental impacts. In this sense, Gursel et al. (2016) investigated strategies for reducing the embodied energy and environmental impacts of both cement usage and aggregate consumption in concretes employed in Singapore. In this study, two cases were considered: (i) the use of materials (cement and aggregate) prepared in Singapore and (ii) the use of materials imported from Malaysia. Gursel et al. (2016) concluded that the by using the Malaysia cement and aggregates instead the corresponding Singapore materials, the potential and embodied energy of a high-strength concrete were reduced by 11% and 31%, respectively.

The effects of the usage of by-products and waste materials as supplementary cementitious systems have been extensively investigated (Xu et al. 2019; Szulej et al. 2019; Long et al. 2018; Warati et al. 2019; Kim et al. 2018; Nguyen et al. 2020; Sofi et al. 2019; Chen et al. 2019; Azmee and Shafiq 2019; Pushkar 2019; Singh et al. 2017; Ruan and Unluer 2017; Sakir et al. 2020; Huang et al. 2017). In this sense, the following systems were added to green concretes: fly ashes (Chen et al. 2019; Huang et al. 2017); metakaolin (Elizondo-Martinez et al. 2020); geopolymers (Araya-Letelier et al. 2019; Parra et al. 2019); slags (such as steel slags) (Kim et al. 2018; Nguyen et al. 2020); and sugar cane bagasse ash (Arif et al. 2017; Molin Filho et al. 2019; Moretti et al. 2018; Singh et al. 2018).

In particular, the sugar cane bagasse ash is a by-product resulted from boilers of sugar and alcohol factories (Arif et al. 2017; Molin Filho et al. 2019; Moretti et al. 2018; Singh et al. 2018). This by-product material, composed mainly of silica (SiO_2), can be used as a mineral admixture in mortar and concrete, with great economic and environmental advantages.

This work intent to collaborate for the partial substitution of Portland cement by sugar cane bagasse (SCB), in order to reduce the consume of clinker by volume of concrete, which is the responsible for the high emission of CO_2 to the atmosphere. This material is often disposed of in dumps and has no use. When its incorporation into cement matrices, a new reuse arises, given that bagasse is not the raw material of sugar production and is already used to generate energy, having as tailings ash, which in turn will be reused in the production of concrete and mortars. For this, an experimental campaign was carried out with cementitious pastes to evaluate the changes in their properties, namely the durability, due to the use of SCB ash.

2 Experimental Campaign

In this work, the characterization and analysis of the potential of sugar cane bagasse ash was performed through chemical, physical and mineralogical methods, aiming to evaluate its pozzolanicity through direct and indirect techniques. Soon after, accelerated techniques of environmental aggressiveness were performed, with the objective of verifying the durability of concrete structures with the addition of this material.

2.1 *Materials*

In this work two samples of sugar cane bagasse ash were used:

- SCBA–A: Collected from a wood oven in a pizzeria in the Metropolitan Region of Recife, Brazil. It is used as an energy source, being used as source of burn and heat during the process of producing the pizzas.
- SCBA–B: Collected from a sugar cane mill located in the city of Palmares, Pernambuco, Brazil. The bagasse is burned in the mill for heating the boilers where sugar derivatives are produced.

2.1.1 Materials Characterization (Anhydrous)

To physically, chemically and minerally characterize the materials used in their anhydrous state, several experimental techniques were selected in order to support the stage of rational use and viability of the materials, as: (1) Scanning electron microscopy (SEM) analysis; (2) Particle size distribution; (3) Specific mass; (4) Specific area; (5) Unit mass; (6) Fire loss (FL); and (7) Thermogravimetric analysis (TG-DTG).

Some characteristics and properties of sugarcane bagasse ash (SCBA-A and SCBA-B) were evaluated in order to prove the elements that could be used as an addition in the manufacture of mortars and concrete, with the subsequent verification of the hypothesis of increasing durability. The samples used in the characterization and subsequent moulding analyses of the specimens were obtained from the SCBA-A and SCBA-B (see Fig. 1), dried in a greenhouse at 105 °C, passed through the sieving process for 20 min at the speed of 70 rpm in order to obtain the passing fraction by the opening sieve 0.075 mm.

Fig. 1 Sugarcane bagasse ash **a** SCBA-A and **b** SCBA-B

2.1.2 Scanning Electron Microscopy (SEM) Analysis

Through scanning electron microscopy analysis it was possible to investigate significant changes in the sizes and morphology of the particles of the samples, due to the vast variation of the materials involved and the various manufacturing processes. In addition, microscopy was used in the investigation of phase identification with cement hydration reactions and in the development of concrete microtexture at advanced ages, older than 200 days, which will be treated later.

To evaluate the morphology of the materials used in the research, the scanning electron microscopy (SEM) technique was used, which uses a beam of electrons to sweep the sample surface, causing the emission of secondary electrons that are used in the formation of the images to be analysed. The scanning electron microscope (SEM) consists of an electronic column, sample chamber, electronic vacuum system and imaging system. An electron beam is emitted and focused on a region of the sample, resulting in the emission of these electrons with great scattering of energy, and are subsequently collected and amplified in order to provide an electrical signal. The signal is used to modulate the intensity of the electron beam in the cathode ray tube.

For the construction of a complete three-dimensional image of the surface this beam sweeps a specific area of the sample surface, since electrons are generated from the interaction between the beam and matter.

For the analysis of the morphology of the samples, a TESCAN scanning electron microscope (SEM), MIRA3 model, 10 kV voltage and SE detector were used. The samples were placed in a Stub with a carbon tape for image acquisition, with the use of metallization and magnitude (MAG) ranging from 650x to 25.00kx. The results of the morphology of the samples are presented in Fig. 2.

It is possible to observe, through the images of the MEV with the secondary electron detector, that the samples of (SCBA-A and SCBA-B) present irregular particle

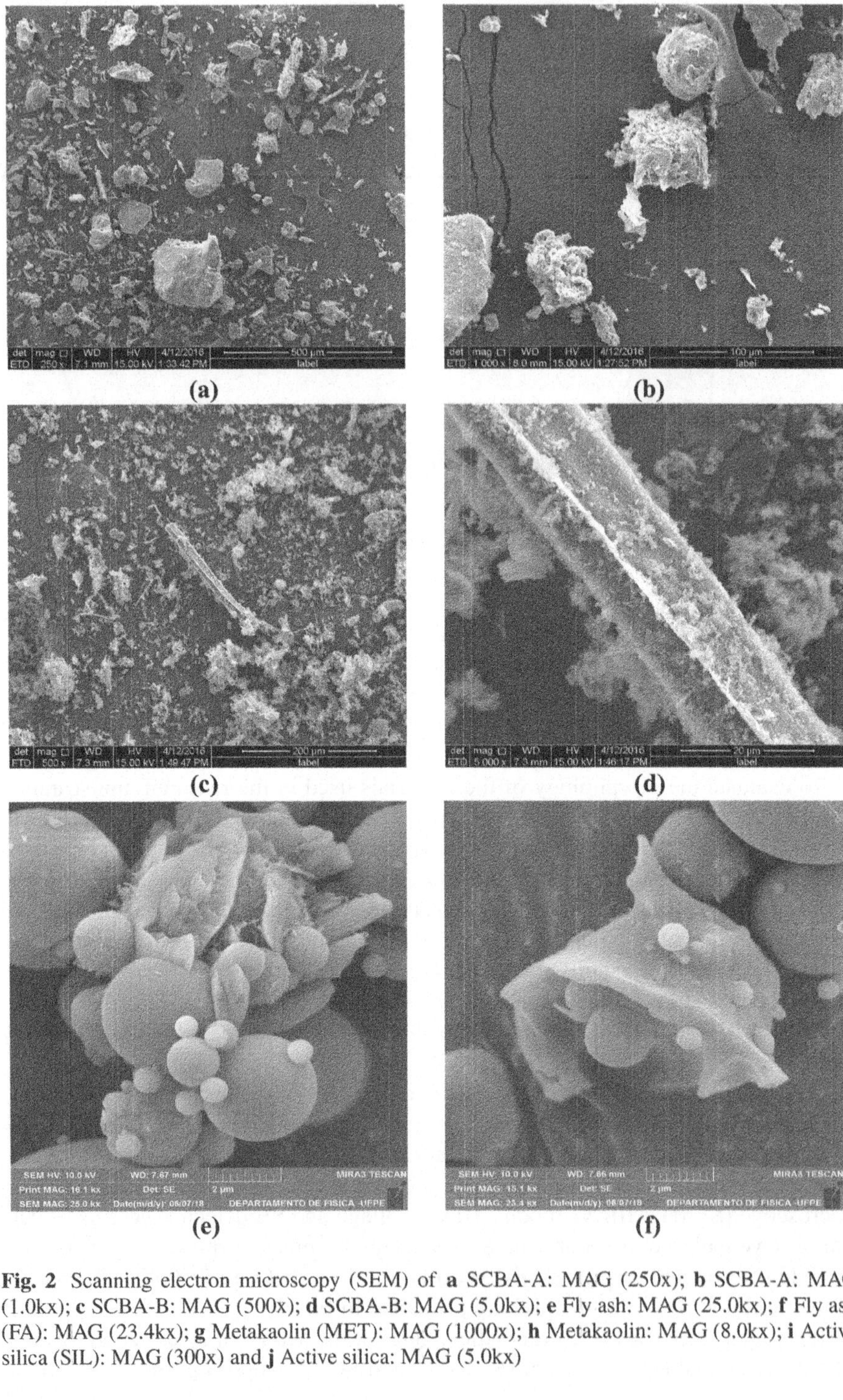

Fig. 2 Scanning electron microscopy (SEM) of **a** SCBA-A: MAG (250x); **b** SCBA-A: MAG (1.0kx); **c** SCBA-B: MAG (500x); **d** SCBA-B: MAG (5.0kx); **e** Fly ash: MAG (25.0kx); **f** Fly ash (FA): MAG (23.4kx); **g** Metakaolin (MET): MAG (1000x); **h** Metakaolin: MAG (8.0kx); **i** Active silica (SIL): MAG (300x) and **j** Active silica: MAG (5.0kx)

Fig. 2 (continued)

forms, rods and lamellar; cavities (caves/voids) are also detected in both samples. It is verified in the sample (FA) the existence of particle clusters and that the particles are spherical and regular. In the samples (MET) and (SIL) are observed quite frequently clusters of particles, because they are very fine materials with a high specific area, but these agglomerates directly influence other tests and results, as well as the irregular particles found in the samples (SCBA-A and SCBA-B).

2.1.3 Particle Size Distribution

For the particle size distribution, the equipment available at the Federal University of Paraíba (UFPB), Cilas 1090 laser granulometer, was used, through the dry and wet route. There are significant differences between the two dry and wet routes, his research warns of care when using the results. According to the method of analysis used, the particle size is given by the diameter of the circumferences that circumscribe them. Thus, if there are rods and particles with long lamellar shapes or particle agglomeration, as visualized in the analysis of scanning electron microscopy (SEM) through secondary electron detectors (SE), the equipment (Cilas 1090 laser granulometer) considers the particle length as diameter, impairing the analysis of this result.

According to this analysis, Portland cement grains (CPV) showed the results for the distributions of medium particles, diameter at 50% (D_{50}) equal to 5.90 μm, while SCBA-A, SCBA-B, FA, MET and SIL presented the average diameter equal to 8.25 μm, 10.32 μm, 9.76 μm, 5.72 μm and 63.67 μm, respectively. In view of the results, the scanning electron microscopy technique was used to ascertain the shape and values of the particles found in the granulometry.

2.1.4 Specific Mass

The specific masses of Portland cement (CPV), metakaolin (MET), active silica (SIL), fly ash (FA) and sugarcane bagasse ash (SCBA-A and SCBA-B) were determined by means of the helium gas pycnometer method, (recommended by ABNT NBR 16605 (ABNT 1660)) using the Accupyc 1330 V2.01—Micrometrics device.

First, the samples are placed in the appropriate container for the test, as exemplified in Fig. 3. Then, the capsule was placed in the equipment for analysis, as illustrated in Fig. 4 and presented in Table 1.

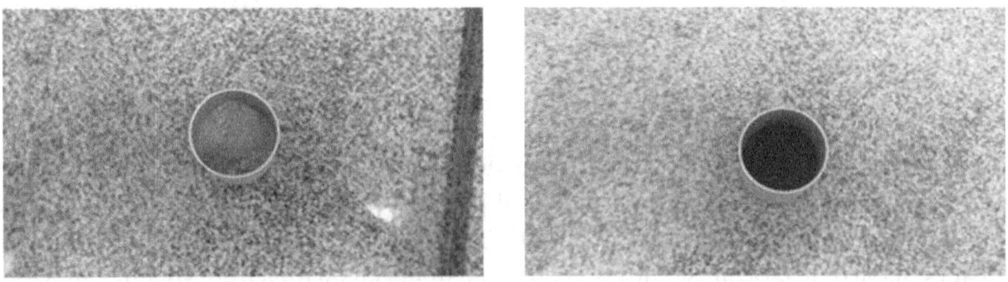

(a) (b)

Fig. 3 Sugarcane ash sample: **a** SCBA-A and **b** SCBA-B

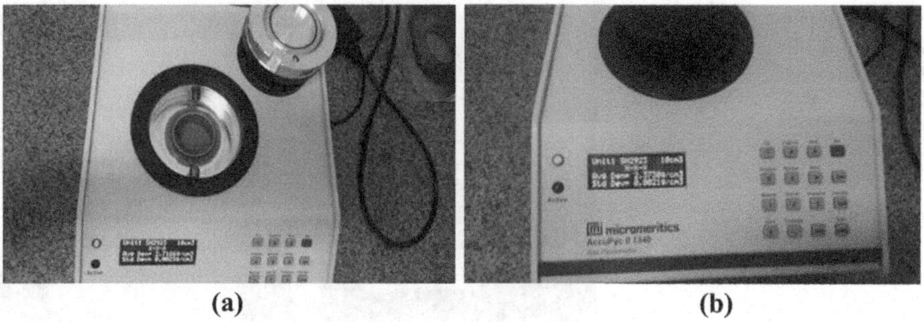

Fig. 4 Sugarcane ash sample for analysis by helium gas pycnometer method: **a** SCBA-A and **b** SCBA-B

Table 1 Physical characterization of materials, particle distribution and fire loss

Properties	CP-V	SCBA-A	SCBA-B	FA	MET	SIL
Unit mass (kg/dm^3)	1.00	0.075	0.089	0.693	0.703	0.550
Specific mass (kg/dm^3)	3.17	2.72	2.37	2.39	2.81	2.22
Specific surface area (m^2/g)	0.452	0.655	0.654	0.678	0.474	0.718
Specific surface area (m^2/g) (BET)	1.784	5.461	22.849	1.158	16.908	20.341
Particle distribution, D_{10} μm	0.94	0.21	0.94	0.18	0.34	2.99
Particle distribution D_{50} μm	5.90	8.25	10.32	9.76	5.72	63.67
Particle distribution, D_{90} μm	29.15	43.95	43.28	92.30	34.36	372.00
Fire loss % (FL)	6.37	19.89	13.64	2.03	3.07	3.09

2.1.5 Specific Surface Area

The specific surfaces of the materials were determined using the Blaine Semi-Automatic ACMEL BSA 1 Permeabilmeter equipment (see Fig. 5), with connection to microcomputer through its serial output RS 232, which expresses the results in cm^2/g on its display, noted and expressed in Table 1. The test takes place without operator intervention. According to standards: NBR 16372 (2020)—Blaine permeabilimeter semi-automatic 220 V-60 Hz 4.257.225.

2.1.6 Specific Surface Area (BET)

The tests of the specific areas per unit of mass were performed using the Brunauer, Emmett and Teller (BET) method on the materials used in the research. The determination of this parameter is important, because materials of the same particle size may present different surface areas due to the inner voids, the shape and roughness of the particles (Cordeiro 2006). Being expressed in square meters per gram, which

Fig. 5 Test using the Blaine Semi-automatic Permeabilmeter

according to Ribeiro and Morelli (2014) is fundamental for the evaluation of material reactivity, the technique consists in measuring the changes suffered by a mixture of nitrogen (N_2) and helium (He), when a sample is subjected to pressure variation in the temperature of liquid N_2. The mathematical equations based on the works of Brunauer, Emmet and Teller (BET) allows calculating the specific area of the material, from the physical adsorption of N_2 molecules.

Checks of the specific surface areas bet of the materials were carried out with the Micrometrics BRAND ASPA 2420 Surface Area and Porosity Analyzer device. As illustrated in Fig. 6, the results were taken note from the software integrated with the equipment and are presented in Table 1.

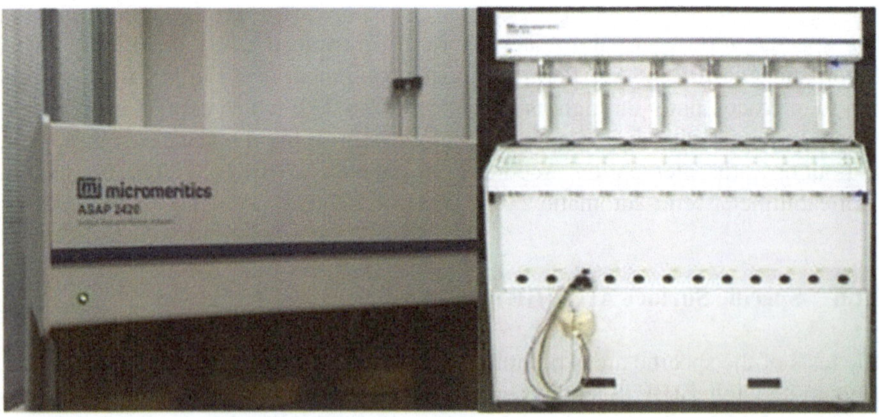

Fig. 6 Testing of the materials used in the research, verification of the specific area

Fig. 7 Fire loss test in muffle furnace

2.1.7 Unit Mass

The unit masses of the materials used were tested according to NM 45 (NBR NM 45 2006)—Determination of the unit mass.

2.1.8 Fire Loss (FL)

This test aims to determine the content of organic materials and other volatile total solids in the materials used in the research, according to NBR NM 18 (NBR NM 2012). During the procedure, the percentages of mass loss of the material after burning at high temperature in a muffle furnace are verified (see Fig. 7).

The amount of material remaining corresponds to the amount of inorganics, since the lost mass represents the amount of organic, volatile and moisture of the sample. In this test a certain mass of material is dried in an oven and after mass constancy, it is subjected to calcination at a temperature of 1000 °C in the muffle equipment. The results are described in Table 1.

The results of the tests carried out, such as particle distributions, determinations of specific masses, specific or surface areas specific to materials, unit masses of materials and loss to fire, are shown in Table 1.

2.1.9 Thermogravimetric Analysis (TG-DTG)

Thermogravimetry is a technique that is based on the measurement as a function of the temperature of the mass variation of samples subjected to a temperature range in

Table 2 Thermogravimetric technique (TG) mass loss (30–1000 °C)

Properties	CP-V	SCBA-A	SCBA-B	FA	MET	SIL
Mass loss % (TG)	6.08	22.22	11.65	1.34	2.63	2.98
Fire loss % (FL)	6.37	19.89	13.64	2.03	3.07	3.09

predefined heating. Essentially, thermogravimetry equipment consists of microbalance coupled to the resistive furnace. The material, in this case, is characterized through thermogravimetric curves, which relate temperature to mass loss over time. To validate the results of fire loss obtained in a muffle furnace, thermogravimetric tests were used.

The analyses were conducted in STA 449 F3 JUPITER NETZSCH simultaneous analysis equipment. The experimental conditions used were: inert atmosphere (N_2), maximum flow of 100 ml/min and heating range of 30–1000 °C, with a heating rate of 10 °C/min, with sample port (alumina crucible), mass approximately around 20 mg ± 0.5 and purge of 50 °C/min. Results are expressed in Table 2.

To verify the accuracy of the estimate of the fire loss test according to NBR NM 18 (NBR NM 18 2012), the results obtained for the samples of the materials were compared with the mass loss obtained by thermogravimetry. Table 1 shows the values of the six fire loss determinations, 6.37%, 19.89%, 13.64%, 2.03%, 3.07 and 3.09% for the samples (CP-V, SCBA-A, SCBA-B, FA, MET and SIL), respectively.

By the thermogravimetry test, the mass loss at the approximate temperature of 1000 °C was 6.08%, 22.22%, 11.65%, 1.34, 2.63 and 2.98%, respectively. In this case, the difference between the values obtained by the two techniques (see Table 2) is only 0.29%, 2.33%, 1.99%, 0.69%, 0.44% and 0.11% for each sample, respectively. It is noticeable that the fire loss test result is significantly accurate, and this analysis corroborates the conclusions reached by Cordeiro (2006).

2.2 *Evaluation of the Pozzolanicity*

2.2.1 Method for Pozolanicity Assessment (Direct)

Through the direct methods, the pozzolanicity of the materials proposed in the research was analyzed. For the identification of the hydrated phases formed and possible consumption of other phases (Portlandite phase), portland cement-based pastes were made without adding materials (reference) and with addition (15%) of the materials studied. The proportions of the materials used in the production of the samples for this analysis (hydrated phases) are presented in Table 3.

According Taylor (1998), these observations in the formation of phases, as well as observation of the consumption of others, follow as a parameter for the hypothesis of possible pozzolanic reactions of the added materials, where such reactions are qualitatively evaluated.

Table 3 Proportion of materials used in the production of pastes

Materials	Mixtures			
	Cement (g)	Water (g)	Added materials (g)	Water/binders
Reference	20	9.60	–	0.48
SCBA-A	20	11.04	3	0.48
SCBA-B	20	11.04	3	0.48
FA	20	11.04	3	0.48
MET	20	11.04	3	0.48
SIL	20	11.04	3	0.48

To prepare the samples (pastes), the remaining 2/3 water was mixed manually in a mechanical mixer and added to a mechanical mixer for a time interval of 1 min and a half. The samples after preparation were added to the moulds and after 24 h were demoulded and added to the submerged cure properly identified, according to Fig. 8a, b.

According to Taylor (1998), for crystallographic analysis, for the specific case of the research, it is recommended that the analysed material be dry and in the appropriate minimum granulometry, i.e., sieve #200 (0.075 mm). Thus, this technique and recommendation were used in the analysis of the pastes, using the powder method according to Cordeiro et al. (2009). In the preparation of the samples for the test itself, the samples are crushed with mortar and porcelain pistil (Fig. 9), and an aliquot of the sample is used in approximately 1.0 g of the passing material in the 0.075 mm sieve, according to Fig. 10a, b.

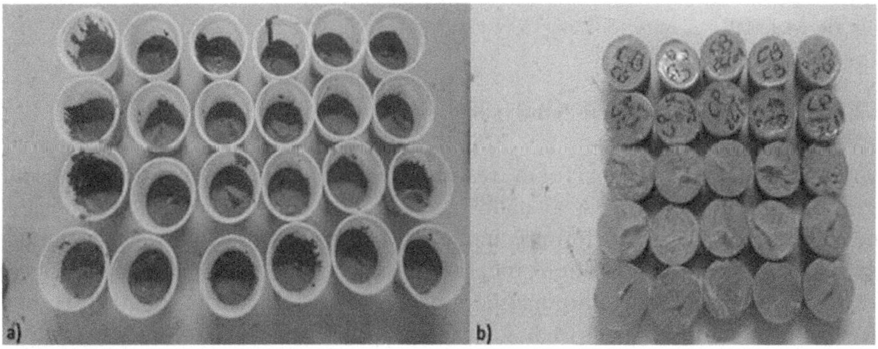

Fig. 8 Demoulding and identification of paste samples

Fig. 9 Crushing of samples with mortar and porcelain pistil

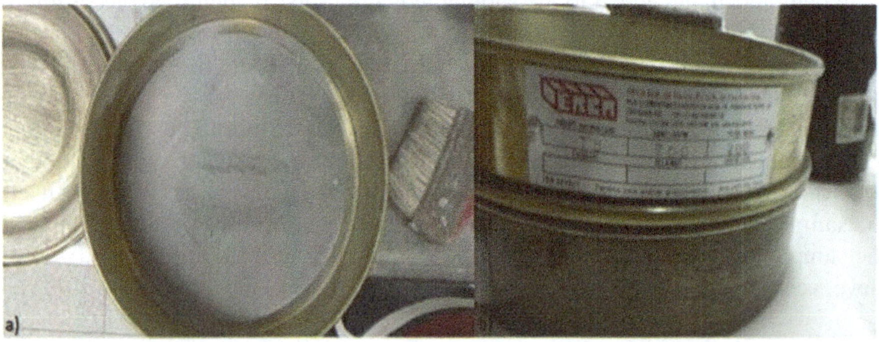

Fig. 10 Material used in the opening sieve #200

2.2.2 Thermogravimetric Analysis

Aiming to support the analysis of the research, the samples were submitted to thermogravimetric analyses with 28, 63, and 91 days of age, being observed the influence of the additions of 15% in the formation of phases of the cementmarices. The hydrated products formed in the hydration process of Portland cement and possible pozzolanic reactions can be identified through thermoanalysis techniques through their respective chemical compositions (Mehta and Monteiro 2014). Thus, in order to evaluate the influence of the use of additions on cement hydration, thermal analyses were used, which allowed identifying the chemically combined water content of hydrates and the possible consumption of portlandite, through pozzolanic reactions.

For thermogravimetric analysis in the specific case of the research, it is recommended that the analysed material be dry and in the appropriate minimum granulometry, sieve #200 (0.075 mm). The analyses were conducted in STA 449 F3

JUPITER NETZSCH simultaneous analysis equipment. The experimental conditions used were: inert atmosphere (N_2), maximum flow of 100 ml/min, initial heating range of 30–1000 °C, with a heating rate of 10 °C/min, mass approximately around 20 mg ± 0.5 and purge of 50 °C/min. From Eqs. (1)–(6), the chemically combined water content (AQC) related to the formation of C-S-H and hydrated aluminate phases, calcium hydroxide $Ca(OH)_2$ and calcium carbonate content ($CaCO_3$) were calculated.

$$CH_1 = \frac{(H_2O \times 100)}{A} \tag{1}$$

$$CO_2 = \frac{(PCO_2 \times 100)}{B} \tag{2}$$

$$C_aCO_{3real} = CO_2 - Approx. \tag{3}$$

$$CH_2 = \frac{(C_aCO_{3real} \times C)}{100} \tag{4}$$

$$AQC = AQC_{total} - CH \tag{5}$$

$$Ca(OH)_2 = CH_1 + CH_2 \tag{6}$$

where H_2O is the water loss that happens in the thermal event of portlandite decomposition, A is the molar water mass H_2O, PCO_2 is the water loss that happens in the thermal decarbonation event, B is the molar mass of the CO_2, Approx. is the 5% approximately the carbonate content present in the CP-V ARI, C is the molar mass of the $Ca(OH)_2$, AQC_{total} is the mass loss occurring in the thermal event ranging from 100 to 600 °C, $CH_1 = CH_2$ is the mass loss that occurs in the thermal event ranging from 350 to 450 °C, AQC is the chemically combined water, $Ca(OH)_2$ is the total portlandite content, and $CaCO_3$ is the calcium carbonate content.

2.2.3 Determination of Pozzolanicity by the Modified Chapelle Method

According to NBR 15895 (ABNT 2010), to perform the method, first 1 g of the material that is intended to evaluate "pozzolanic material" (m_2) and 2 g of calcium oxide (m_3) should be weighed. Immediately thereafter, 250 ml of CO_2-free distilled water should be added, measured in a volumetric flask in an Erlemeyer heated to 90 ± 5 °C, stirring in a container simultaneously for 16 h (see Fig. 11).

In parallel to the test with the pozzolanic material, the same procedures are performed for the determination of the blank test containing only calcium oxide, under the same conditions. This result is used as a starting point for the actual amount of Ca^{2+} available in solution, as it will discount in titration the amount of material that precipitates under such conditions.

Fig. 11 Samples tested—Digital water bath (CienlaB)

In a third moment, 250 ml of sucrose solution was added to Erlemeyer with material evaluated "pozzolanic material" and calcium oxide, which was shaken for 15 min. It was pipetted 50 ml of this solution and titrated with 0.1 M chlorine hydroxide solution and using 1 g/l of phenolphthalein solution as an indicator. The volume consumed in the titration was noted (V_2). The same procedure was performed with the blank test, noting the volume spent (V_3).

The pozzolanic activity index by the modified Chapelle method can be calculated according to Eq. (7), corresponding to the amount of calcium hydroxide used (in mg) per gram of pozzolanic material.

$$I_{Ca(OH)_2} = \frac{28x(V_3 - V_2) \times Fc}{m_2} \quad (7)$$

where $ICa(OH)_2$ is the pozzolanic activity index corresponding to the fixed calcium hydroxide content (mL), M_2 is the mass of pozzolanic material (g), V_2 is the volume of HCl 0.1 M consumed test with pozzolan solution (mL), V_3 is the volume of HCl 0.1 M consumed in the test with reference solution (mL) and Fc is the correction factor of the HCl for a concentration of 0.1 M (Fc = 1, 32).

This methodology allows to determine the pozzolanicity of the materials used as mineral addition as a function of the reaction rate, after a predetermined time, and the reaction is accelerated by temperature rise, that is, a certain amount of supposedly pozzolanic material and CaO are placed to react in a water bath at approximately 90 ± 5 °C for 16 h. According to Raverdy et al. (1980), the material is considered pozzolanic if the calcium oxide intake is greater than 330 mg CaO/g, a value that by stoichiometry is equivalent to 436 mg of $Ca(OH)_2$ per gram of pozzolan. In the present research, this method was applied to measure chemical reactivity, and portlandite consumption was evaluated by SCBA-A and SCBA-B. In addition, active silica (SIL), metakaolin (MET) and fly ash (FA) were also analysed.

2.3 *Method for Pozolanicity Assessment (Indirect)*

Through indirect methods, the pozzolanicity of the materials proposed in the research was analysed.

2.3.1 Electrical Conductivity

For the analysis by electrical conductivity, in the verification of the possibilities of the pozzolanic activities of the studied materials, the technique is defined by the loss of electrical conductivity of a solution of calcium hydroxide P.A, based on the research developed by Payá et al. (2001), Lúxan et al. (1989).

A 1000-second analysis period was defined, proposed by the authors Payá et al. (2001) who observed in their research as being sufficient to verify the pozzolanic activity of sewage sludge ash, among other supplementary materials. Thus, this time interval was fixed as being sufficient to analyse the ashes of sugarcane bagasse. The test itself was carried out with the following materials and procedures described in:

- A 250 ml glass beaker, a rubber stopper (in an attempt to prevent carbonation of the solution, according to the authors Payá et al. (2001) (see Fig. 12a);
- Magnetic stirrer, which allows the adjustment of the base plate temperature and stirring rate (see Fig. 12b);
- A Digimed DM-32 v.2.0 conductivimeter, which has an RS-232 output and a Digimed DMC-001 XTX conductivity cell used for the conductivity measurements of the solution (see Fig. 12a).

The rubber stopper has three holes, which are designated: the conductivity cell, the temperature controller thermometer of the plate and the tube for sample insertion. For the preparation of calcium hydroxide solution, the guidelines described by Payá et al. (2001) were followed, using 800 mg of $Ca(OH)_2$ per liter of deionized water. It is important to use deionized water so as not to compromise the results, being observed by Payá et al. (2001) such influences on the results in their research.

Fig. 12 **a** 250 ml glass beaker and a rubber stopper and Digimed DM-32 Condutivmeter; **b** Magnetic stirrer with temperature settings

Calcium hydroxide was dissolved in water in a glass container. At first, the initial preparations were made with 1050 ml of distilled water heated to 60 ± 2 °C, and this temperature was determined due to the similarity with the initial temperature of Portland cement hydration (Mehta and Monteiro 2014). After the water reached the ideal temperature and its desired stabilization, 840 mg of $Ca(OH)_2$ P.A. (minimum of 95% purity) was added. The container was then sealed and the solution was kept under constant agitation at 700 rpm for a period of 1 h.

After this initial preparation period, 200 ml of this solution was used, transferring it to the beaker with a capacity of 250 ml under the same conditions previously established and the rubber stopper was inserted firmly, together with the conductivity cell and thermometer; followed by an addition of 4 grams of the materials studied in this research (SCBA-A, SCBA-B, FA, MET and SIL) in $Ca(OH)_2$ solution, through the appropriate tube fixed to the stopper, according to the proportions established in the study by Payá et al. (2001).

The measured results of electrical conductivity are expressed in mS/cm and were collected through the Raspberry Pi every second, from the immediate moment of contact of the sample with the solution until the end of the established time, that is, end of the 1000 s of the test. Finally, the curves of variation of electrical conductivity and the loss of this conductivity over time are obtained, the data obtained from a variation of electrical conductivity for a given period of time (−t) and a relative loss for a given time $(\%PR_{Cond})_t$, should be calculated according to the Eq. (8) presented in Payá et al. (2001).

2.3.2 Portland Cement Performance Index at 28 Days

The performance index with Portland cement in the pozzolanic activity of the materials was determined following part of the physical and chemical requirements proposed by NBR 12653 (ABNT NBR 2020). According to the above standard, there are the minimum physical requirements for a material to be considered pozzolanic: a minimum of 90% performance index with Portland cement in relation to the control.

The pozzolanic reaction of the materials used in this research was evaluated by determining the performance index with Portland cement according to NBR 5752 (ABNT NBR 2014), for this purpose two different mortar compositions suggested by the standard are used: (i) mortar A, containing only Portland CP II F-32 cement as binder and (ii) B mortar, containing 25% of the material analysed as pozzolanic (SCBA-A and SCBA-B), replacing cement. The materials used in this test are described in Table 4.

Six cylindrical specimens (5 cm × 10 cm) of mortar moulded for each trace were made and, after 28 days, these specimens were ruptured by axial compression according to NBR 5739 (ABNT NBR 2018).

Table 4 Determination of pozzolanic activity index with cement

Materials	Mass (g)			
	Mortar A	Mortar B (SCBA-A)	Mortar B (SCBA-B)	Water/binder
CP II F-32	624 ± 0.4	468 ± 0.4	468 ± 0.4	0.48
Pozzolanic material	–	156 ± 0.2	156 ± 0.2	0.48
Sand normalized by IPT NBR 7214 (ABNT NBR 2015)	1872.0	1872.0	1872.0	–
Water	300 ± 0.2	300 ± 0.2	300 ± 0.2	–

2.4 *Analysis of the Durability of Concrete*

Through the durability methods, the additions of the materials proposed in the research were analysed.

2.4.1 Moulding of Specimens 5 cm x 10 cm (Mortar)

For the evaluation and analysis of the durability of concrete follow the tests at different ages of submerged curing: accelerated carbonation, total absorption, capillary and immersion. All with the objective of analysing the efficiency of the addition of mineral materials. Cylindrical specimens of 5 cm × 10 cm of mortar were moulded according to NBR 7215 (ABNT NBR 2019). Using a normal sand described in NBR 7214 (ABNT NBR 2015), the ash of sugarcane bagasse from the pizzeria (SCBA-A), the ash of the sugarcane bagasse from the plant (SCBA-B), the fly ash (FA), the metacaulim (MET), the active silica (SIL) and the Portland CPV-ARI cement (REF).

The mixture used in the preparation of 5 cm × 10 cm specimens was (1:3:0.48), with the materials described in Table 5 and presented in Fig. 13. Regarding the addition levels, 15% addition to cement was used for the materials used in the research (SCBA-A, SCBA-B, FA, MET and SIL).

Table 5 Material used for the mixture

15% content	Cement (g)	Sand (kg)	Water (ml)	Mineral (g)
REF	624	1872	300	–
SCBA-A	624	1872	300	93.6
SCBA-B	624	1872	300	93.6
FA	624	1872	300	93.6
MET	624	1872	300	93.6
SIL	624	1872	300	93.6

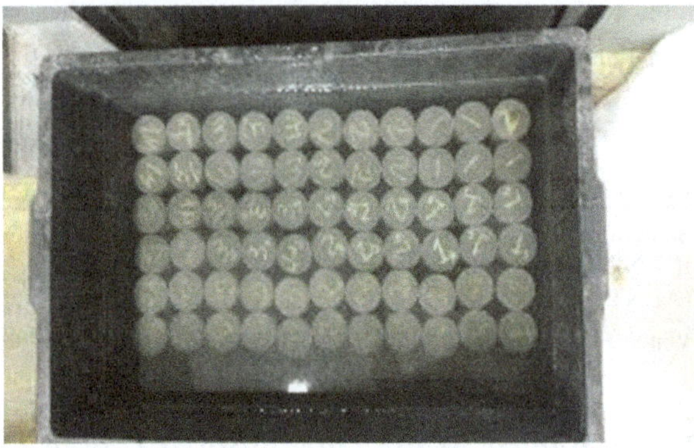

Fig. 13 Submerged healing of mortar specimens

Table 6 Total number of samples for each test

Samples	Accelerated carbonation	Capillary and immersion absorption
REF	06	05
SCBA-A	–	–
SCBA-B	06	05
FA	06	05
MET	06	05
SIL	06	05
Total	30	25

Twenty-four more (twenty-four) specimens were made with the addition of 15% (SCBA-B, FA, MET and SIL) and 6 (six) specimens without 0% addition (references); totalling 30 (thirty) specimens for the accelerated carbonation test.

Finally, 20 (twenty) more specimens were made with the addition of 15% and 5 (six) specimens without adding 0% (references); totalling 25 (twenty-five) specimens for the absorption, capillary and immersion test. The total number of samples and their respective tests are described in Table 6.

2.4.2 Moulding of Specimens 10 cm × 20 cm (Concrete)

For the evaluation and analysis of the durability of the concrete following with the objective of analysing the efficiency of the addition of mineral materials in Portland cement, cylindrical specimens of 10 cm × 20 cm of concrete were moulded according to NBR 5738 (ABNT NBR 2015). The specimens used medium sand, gravel nr. 19, water supplied by the Pernambuco Sanitation Company (COMPESA), sugarcane bagasse ash (SCBA), fly ash (FA), metakaolin (MET), active silica (SIL) and Portland cement CPV-ARI (REF).

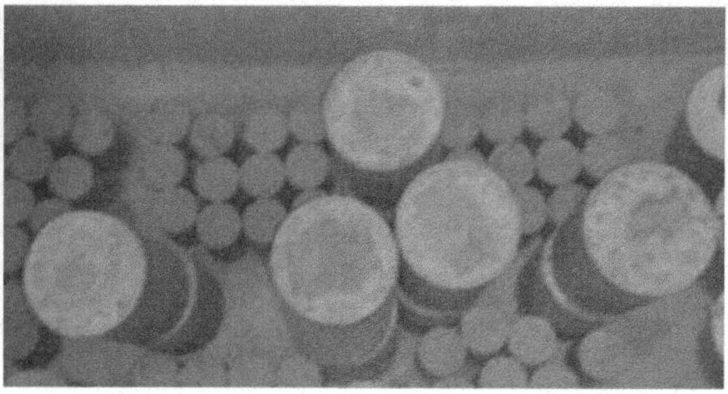

Fig. 14 Submerged healing of concrete specimens

The mixture used in the preparation of the specimens 10 cm × 20 cm cm was (1:1.51:2.06:0.48) in which we have cement, sand, gravel and the water/cement ratio, with 45% gravel, A% = 10.50, α = 0.55 and cement consumption (kg/m^3) = 465.61, for the preparation of 2 (two) specimens in each of the strokes with additions and without addition with 0% called reference, totalling a sample space of 10 (ten) test cups, for the accelerated migration test of chlorides (see Fig. 14).

2.4.3 Total Absorption, Capillary and Immersion

The durability of concrete is a very important factor to ensure the integrity of a concrete structure during its useful life and vital is the mechanical aspect with its compressive and tensile strengths. For the American Concrete Institute, through the 201 Committee, the durability of concrete can be defined as the ability to resist environmental actions, chemical attacks, abrasion and any other deterioration process (ACI 201.2R-01), corroborating the provisions of NBR 6118 (ABNT NBR 2014). Thus, the study of the interaction between the interface of the concrete surface layer with the external environment is determinant for the understanding of degradation processes through the ingress of CO_2 chlorides with carbonation among others and through wetting drying cycles. These processes, in general, are associated with physicochemical phenomena between concrete constituents and aggressive agents in the presence of water. Therefore, the evaluation of total capillary absorption and immersion and void index of concrete become relevant in estimating the durability of a structure constructed with this material.

2.4.4 Water Absorption Test by Capillarity

The execution of the water absorption test by capillarity followed what the procedures and recommendations of NBR 9779 (ABNT NBR 2012). This standard provides for measurements of the mass and height of water absorption in mortars. The speed of capillary ascent is monitored by mass gain, being a phenomenon that occurs in porous structures and consists of the action of the forces of attraction of the pores

Fig. 15 Support and water slide specimens in capillarity test

of the structure on the liquids that are in contact with the surface (Cordeiro 2006; Neville 2015). After 28 days of submerged curing, the mortar specimens were packed in an oven with a temperature of 45 °C ± 2 °C, with internal ventilation until mass constancy for 23 days, remaining to perform the test. A constant temperature of 45 °C ± 2 °C was maintained in order to maintain the integrity of Portland cement hydration products, according to Cordeiro (2006).

In a container installed in an environment with a constant temperature of 23 ± 2 °C, the specimens were positioned on a metal support (see Fig. 15), filling the container with water at a constant level of 5 mm ± 1 mm above the lower face. The saturated mass was determined with 3 h, 6 h, 24 h, 48 h and 72 h. Previously the excess water was removed with damp cloth and after each step returning them to the container with the constant water slide.

The test was performed with three cylindrical mortar specimens 5 cm × 10 cm, for each sample analysed and the water absorption by capillarity is calculated according to Eq. (8).

$$C = \frac{Msat - Ms}{S} \tag{8}$$

where C is the absorption of water by capillarity (g/cm^2); $Msat$ is the saturated mass of the specimen that remains with one of the faces in contact with water for the specified period of time (g); Ms is the mass of the dry specimen, as soon as it reaches the temperature of 23 ± 2 °C (g) and S is the cross-sectional area (cm^2).

2.4.5 Immersion Water Absorption Test (A(%))

The conduction of aggressive agents such as chlorides, CO_2, among others, occurs through absorption and permeability, and this conduction is governed by the rate of water penetration, because only through moisture are the agents transported to the interior of the concrete. Thus, the NBR 9778 (ABNT NBR 2009) standard was used as a parameter to evaluate the absorption capacity of water by immersion, which classifies as the phenomenon by which water is transported tending to occupy the permeable pores of a porous solid body and also as the increase in mass of a porous solid body due to the penetration of water into its permeable pores until the moment of saturation, in relation to its dry mass.

According to NBR 9778 (ABNT NBR 2009), the samples were taken to the greenhouse for 72 h at a temperature of 45 °C ± 2 °C, with internal ventilation until mass constancy for 23 days, remaining for later the test with a temperature of 45 °C ± 2 °C, in order to maintain the integrity of Portland cement hydration products., according to Cordeiro (2006). Soon after, the masses of the dry samples were determined and recorded, at the second moment the samples were immersed in water at a temperature of (23 ± 2) °C for 72 h, with the aid of a hydrostatic scale, the mass of the immersed sample was determined. The test was performed with three cylindrical mortar specimens 5 cm × 10 cm, for each sample and the water absorption by immersion is calculated according to Eq. (9).

$$A_{\%} = \frac{Msat - Ms}{Ms} \times 100 \tag{9}$$

where Msat is the mass of the saturated sample after immersion and Ms is the mass of the sample dried.

2.4.6 Determination of Void Indexes (Iv(%))

The total water absorption test, according to NBR 9778 (ABNT NBR 2009) enables and enables the calculation of the void index Iv(%) cylindrical specimens. This index is associated with the total volume of pores accessible to water and does not reflect the ease with which water can enter the concrete.

Thus, the void indexes of the samples were measured in analyses, according to NBR 9779 (ABNT NBR 2012). The mean values calculated in the void index test were performed on three cylindrical mortar specimens 5 cm × 10 cm, for each sample and the determination of the void indices were performed according to Eq. (10).

$$I_{v(\%)} = \frac{Msat - Ms}{Msat - Mi} \times 100 \tag{10}$$

where *Mi* is the mass of the saturated sample immersed in water; *Msat* is the mass of the saturated sample after immersion and *Ms* is the mass of the sample dried in greenhouse.

2.4.7 Determination of Specific Mass

In addition to the void index, the absorption test allows the establishment of the specific mass that is determined by the relationship between the mass and volume of a given material. According to NBR 9778 (ABNT NBR 2009), the specific masses of the dried samples, specific masses of the saturated samples and the actual specific mass can be determined, depending on the status of the samples when they are weighed (dry or saturated), as well as the volume used, total volume or volume occupied by the solid material.

Thus, according to NBR 9778 (ABNT NBR 2009) the specific dry (ρs), saturated (ρsat) and actual (ρr) masses of the samples were determined for each sample and according to Eqs. (11)–(13), respectively.

$$\rho_s = \frac{Ms}{Msat - Mi} \tag{11}$$

$$\rho_{sat} = \frac{Msat}{Msat - Mi} \tag{12}$$

$$\rho_r = \frac{Ms}{Ms - Mi} \tag{13}$$

2.4.8 Resistance to Accelerated Carbonation

According to Cascudo et al. (2014), in terms of durability of concrete structures, carbonation is one of the pathological manifestations that negatively affects the durability and useful life of structures, because the consequences of these manifestations are: depassivation and corrosion of reinforcements, with the exception of simple concrete. For Helene and Ribeiro (n.d.), carbonation of concrete are relatively slow reactions in natural conditions. On the other hand, Medeiros (2002) warns of the concern to obtain results in a short period of time when special interest is observed in accelerated carbonation testing methods using high concentrations of CO_2. With these methods it becomes possible to reach carbonation depths equivalent to many years in a few weeks. Medeiros (2002) also warns that the lack of a standardised existing method is a negative point, but currently in Europe there is concern about the development of standards aimed at this aspect.

On the aspect of CO_2 concentration used in the tests, Dunster (2000) studied a variety of concretes submitted to various CO_2 concentration levels (1%, 4% and 10% CO_2) and found that there was a significant increase in the carbonation front with an increase from 1% to 4% in CO_2 content. However, when it exceeded from 4% to 10% the increase was not significant. The exception is the cases where large amounts of fly ash and blast furnace abrasion were used, i.e. 40% and 60% in relation to the cement mass, respectively (Dunster 2000).

According to Dunste (2000), when the CO_2 concentration varied between 7 and 18%, the carbonation rate was not significantly affected for concretes with compressive strength above 40 MPa at 28 days. Thus, this research fits perfectly in a concentration below 7%, because the average resistance in which the specimens were dosed and molded are above 40 MPa. Therefore, the choice of 5% CO_2 concentration for this research is perfectly adequate to the studies prepared by Medeiros (2002).

In this research the carbonation camera used is a CARON—Refrigerated CO_2 Incubator and has automatic digital control of temperature ± 1 °C, relative humidity ± 2% and controlled CO_2 content, according to Fig. 16a. Initially the samples were placed in the dry chamber until they reached the constancy of masses according to Fig. 16b and remained in this environment for 19 days. Once the mass constancy was reached, the bodies were extracted from the dry chamber and placed in the carbonation chamber, where they remained for 30 days, according to Fig. 16c.

The specimens were fractured in three parts each, at each opportunity the carbonation front verification process was carried out in slices with about 3.3 cm of the specimen. For the rupture of the specimens, an EMIC press with a maximum capacity of 100 kN was used.

The carbonation depth was then measured using a 0.01 mm precision caliper. The temperature in the carbonation chamber was kept constant at 23 °C, the relative humidity of 65% and CO_2 content of 5%, and everything is more constant as shown in Fig. 17a. On each regular time, 6 readings were performed in each slice, as shown in Fig. 17b.

For this, a chemical solution of phenolphthalein was used as a chemical indicator of pH. In the preparation of the chemical solution and in the performance of the test, the RILEM CPC-18 (Rilem Recommendations 1988) standard was considered, which stipulates a chemical solution of 1 g of phenolphthalein dissolved in 100 g of 70% ethyl alcohol. The phenolphthalein solution, after its preparation, was stored in a container for later use. The chemical solution used presents colour change in the pH range between 8.3 and 10.0 presenting within it a rosy/crimson red colour.

The technique consisted of the sprinkling of the indicator in the newly fractured region of the concrete. The contact of the solution with the concrete showed the non-carbonated region, where the pH is very alkaline (the typical colouring of the indicator occurred to the pH value of the concrete close to 13), differentiating it

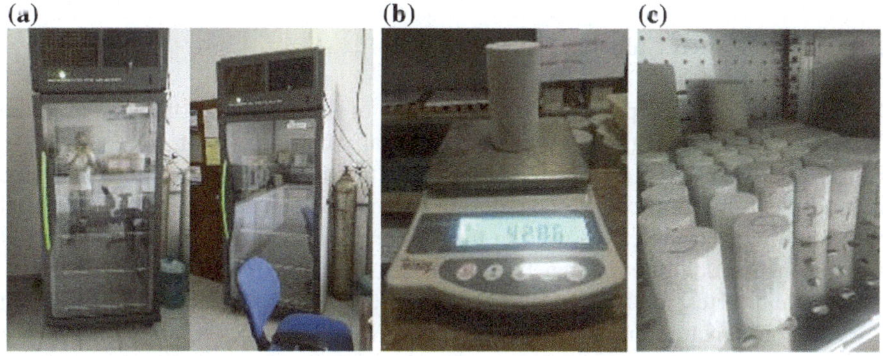

Fig. 16 **a** Carbonation chamber; **b** Measurement for mass constancy and **c** Carbonation chamber with specimens

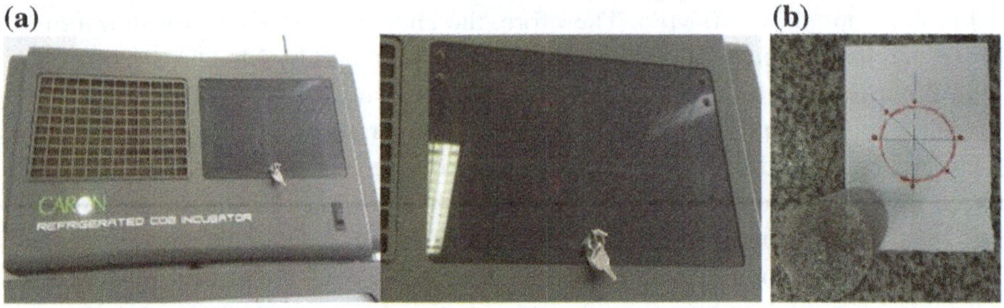

Fig. 17 **a** Parameters for testing in the carbonation chamber and **b** Readings performed on each specimen

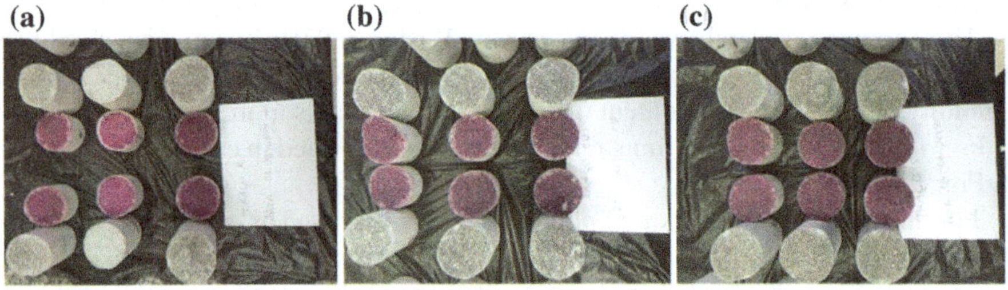

Fig. 18 Samples after testing for carbonation evaluation: **a** reference; **b** Fly ash and **c** active silica

from the carbonated region (which resulted in a colourless aspect, with the pH of the concrete close to 8).

The first rupture was made one month immediately after they were ruptured and the advance of the carbonation front in the concrete was measured. The second rupture was performed after 2 months of confinement of the samples in the carbonation chamber, as showed in Fig. 18.

2.4.9 Accelerated Chloride Ion Migration

Aiming at the research proposal, the evaluation of the behaviour of concrete in the face of the action of aggressive agents, especially chloride, is of great relevance. For, in coastal regions, attacks by this type of aggressive agents are constant and common. For this, the accelerated chloride ion method that determines the electrical conductivity of concrete was used in this research to provide a quick indication of its resistance to chloride ion penetration.

The standard test method for electrical indication of the concrete's ability to resist chloride ion penetration (Cl^-) was based on ASTM C1202 (2020) standard. For the execution of the test and future evaluation of the capacity of concrete to resist the penetration of chloride ions, the aforementioned standard establishes defined standards, which consist of submitting slices of 50 mm thick with 100 mm diameter (Medeiros 2008 and ASTM C1202 2020).

According to Medeiros (2008), the age of the samples has significant effects on the test results, depending on the type of concrete and the curing procedure. Most concrete, when properly cured, becomes progressively and significantly less permeable over time. Therefore, the samples studied aimed at such contributions in the durability of concrete over time when using sugarcane bagasse ash as pozzolanic material. The test was performed at 365 days of age of the samples.

The test is conducted with the application of a voltage of 60 V over 6 h with the measurement of the passing current every 30 min. With the passing current readings, in Amps, over time in seconds, you get a curve. The result of the test is the integral that represents the area between the curve and the axis, obtaining the passing load over 6 h in A/s or Coulombs (C).

First, the specimens were dried in an oven at 100 °C until the mass constancy and packed for 24 h in a laboratory environment, to cool in a dry chamber. In a second moment, protection systems (coating) were applied to one of the circular faces of each specimen and a silicone layer on its sides. The following procedure involved a method of saturation of the specimens by vacuum and immersion. In the cathodic chamber an aqueous solution with 3% NaCl is placed and in the anodic chamber a solution of NaOH 0.3 N.

For the execution of the test, concrete samples (REF, SCBA, FA, MET and SIL) with dimensions of 100 mm in diameter and 50 mm in height and 365 days of submerged curing were positioned between two cells (Fig. 19), negative and positive, containing solutions of sodium chloride (3% by mass) and sodium hydroxide (0.3 N) respectively. They were subjected to a potential difference of 60 V for a period of 6 h.

The voltage applied to a resistor of 0.1 Ω is checked every 30 min. Therefore, the electric current (Ohm's Law) and, consequently, the passing load through the Eq. (14) according to ASTM C1202 (ASTM 1202) is calculated. For each mixture of the different samples the electrical loads were calculated by the arithmetic mean of two determinations in different specimens.

$$Q = K_{\frac{C}{A}}(I_0 + 2I_{30} + 2I_{60} \ldots + 2I_{300} + 2I_{330} + 2I_{360}) \quad (14)$$

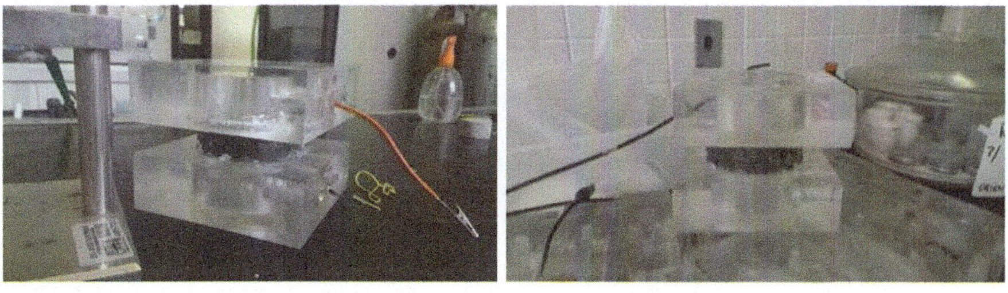

Fig. 19 Cells for chloride migration test and Chloride migration test preparation

where Q is the electric charge; $K_{C/A}$ is the constant equal to 900 C/A and In is the electric current after applying the voltage.

2.4.10 Scanning Electron Microscopy Analysis

Using the use of scanning electron microscopy it is possible to investigate significant changes in the morphology of the particles of the materials used in the research, caused by the different types of samples. In addition, microscopy was used in the investigation of the evolution of hydration reactions and the development of microstructure.

Observing the microanalysis technique, when a small region of the sample is reached by a beam of electrons, different signals are originated, such as secondary electrons and back-spoly electrons, X-ray spectrography, among others Taylor (1998) and Cordeiro (2006). These signals can be detected independently and then transformed into electrical signals, allowing the generation and capture of surface images, phase composition and compositional chemical analysis (Cordeiro 2006). Secondary electrons SE are low energy particles, through the resulting inelastic collision of primary electrons of the beam emitted with electrons from the samples studied. Its recording allows the characterization of the topography of the sample with identification of size, shape and surface texture (Taylor 1998).

According to Taylor (1998), to chemically analyse a specific point or in a sample area in qualitative, semi-quantitative and quantitative terms, energy dispersion X-ray spectrography (EDS) can be used. The methods of qualitative and semi-quantitative analysis are usually available in EDS systems and are usually used because they allow the identification and determination of the composition of the elements present in the sample quickly.

A TESCAN scanning electron microscope, model MIRA3, was used to analyse the morphology of the samples. The main electron beam was generated from tungsten filament with a voltage of 10 kV and SE detector and EDS detector and with working distance of 40 mm.

The samples were glued to a Stub with a carbon tape for image acquisition, with the use of metallization, covered with a thin layer of gold through the metallized layer in order to make them conductive. For point chemical analyses (EDS) a detector of type OXFORD (X-Max^{n}).

3 Results and Discussions

3.1 *Direct Method for Pozolanicity Evaluation and Analysis*

3.1.1 Thermogravimetric Analysis

Studies developed by Dweck et al. (2009) using thermogravimetric analyses in Portland cement pastes, demonstrated that it is possible to determine the chemically combined water content of all hydrated compounds in the thermal event ranging from 50 °C to the final decomposition temperature of portlandite. Thus, analyses were performed through thermoanalysis in reference pastes (without SCBA) and in pastes containing SCBA (SCBA-A and SCBA-B series). These samples were analysed after 28, 63 and 91 days of cure, by thermogravimetry, identifying the hydrated phases formed, as well as observing the possible consumption of Portlandite by reactions with sugarcane bagasse ash. According to Hoppe Filho et al. (2017) it is possible to observe in thermal events the characteristics of each phase.

According to Fig. 20, it is possible to observe the losses of masses occurred in their respective thermal events; especially the highlighted thermal event, happening between 400 °C and 450 °C with the decomposition of portlandite. Thus, thermogravimetry was the analytical technique used in this study to monitor the possible pozzolanic reactions of cement pastes containing 15% addition of ash (SCBA-A and SCBA-B).

Through the ΔH > 0 and exothermic ΔH < 0 endothermic structural transformations, it is possible to perform a quantitative analysis according to Hoppe Filho (2008). Therefore, analyses were made in the pastes at 28, 63 and 91 days through stochiometric calculations and quantifying the contents of hydrated compounds formed,

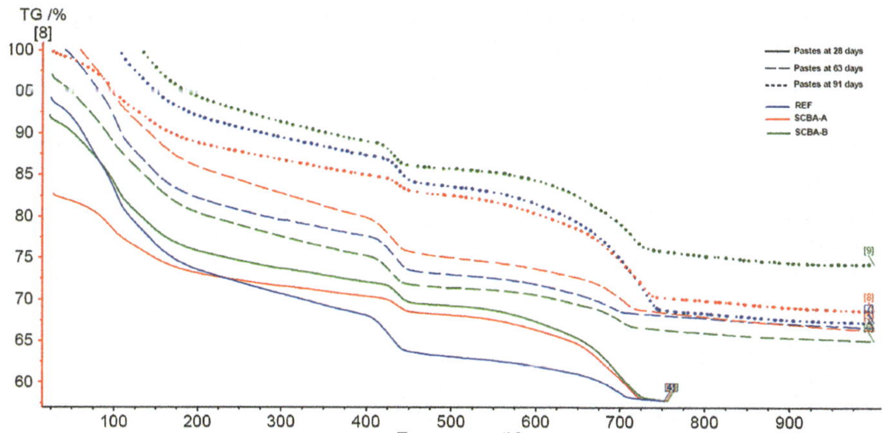

Fig. 20 Thermogravimetric analysis (TG/DTG) of the pastes at 28, 63 and 91 days

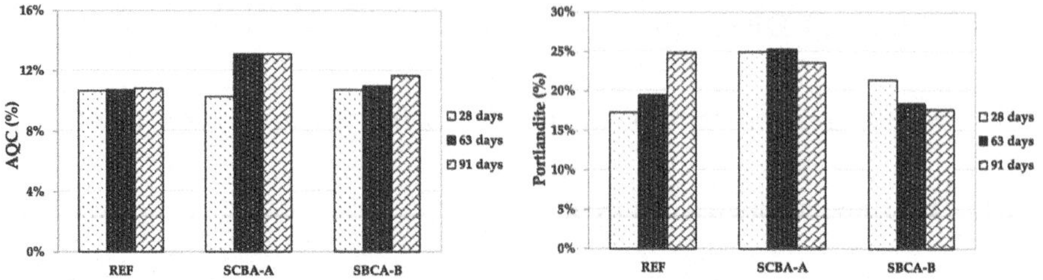

Fig. 21 Chemically combined water content (AQC) and portlandite [$Ca(OH)_2$]

the portlandite content remaining in the hydrated matrix and the calcium carbonate content.

Figure 21 shows the result of thermogravimetric analysis of the pastes, the chemically combined water content (AQC) related to the formation of C-S-H and the hydrated aluminate phases, calcium hydroxide [$Ca(OH)_2$] and calcium carbonate content ($CaCO_3$), which were quantified through Eqs. (1)–(6). Table 7 shows the results after stochiometric calculations of the hydrated compounds analysed.

According to Hoppe Filho et al. (2017), the evolution of cement hydration provides an increasing increase in the portlandite phase [$Ca(OH)_2$], as well as in the chemically combined water content (AQC) due to the hydration of the anhydrous phases C3A, C3S, C2S and C4AF, as noticeable in the reference samples according to Table 7 and Fig. 21. In the analysis of hydrated compounds (NDF) contents for the reference samples, no significant evolution is perceived, as the results remained almost constant. When analysing the event that occurs between 400 and 450 °C, the hypothesis of portlandite decomposition ($Ca(OH)_2$) can be attributed in Fig. 12, since there

Table 7 Summary of chemically combined water and portlandite [AQC and $Ca(OH)_2$]

Cement paste	Cure	AQC (%)	H_2O (%) (mass loss)	$Ca(OH)_2$	CO_2 (%) (mass loss)	$CaCO_{3real}$	Residual mass
Cement CP-V ARI	**28 days**	10.69	3.26	17.28	3.91	2.56	73.46
SCBA-A		10.28	1.89	25.00	10.69	19.30	73.12
SCBA-B		10.73	1.90	21.42	8.66	14.68	72.85
Cement CP V—ARI	**63 days**	10.70	3.61	19.48	4.30	3.31	71.03
SCBA-A		13.11	3.59	25.29	5.26	7.23	71.84
SCBA-B		10.96	2.80	18.37	3.87	3.80	75.74
Cement CP V—ARI	**91 days**	10.83	2.49	24.86	11.75	17.70	68.10
SCBA-A		13.11	1.69	23.54	10.62	24.14	68.59
SCBA-B		11.67	1.85	17.66	6.59	9.97	72.66

was an evolution in the increase of portlandite for reference samples (CPV-ARI). For the analysis of calcium carbonate content in its respective event, at 91 days there was a significant increase described in Table 7 demonstrating that the hypothesis of carbonation of the sample may have occurred, being attributed this fact to other generating sources. In the analyses of the contents of hydrated compounds (AQC) for the samples (SCBA-A and SCBA-B) an increase in these compounds is observed over time described in Fig. 21, in a discrete but quite noticeable way. Regarding the portlandite content [$Ca(OH)_2$], the samples containing 15% ash became quite expressive in quantitative terms for both samples. Thus, it can suggest the hypothesis that the ashes had as main action the nucleation, being a source that generates portlandite. On the other hand, if the evolution of (AQC) and the consumption of [$Ca(OH)_2$] is observed, it becomes noticeable according to Fig. 21. This trend suggests the hypothesis that there were pozzolanic reactions of the samples in the paste.

Based on these results, it can be assumed that the ashes (SCBA-A and SCBA-B) may have acted through the pozzolanic reaction in the refinement of the microstructure, due to the fact that the secondary C-S-H crystals decrease the average pore diameter of the cement matrix as shown by Cordeiro (2006), Duart (2008), Cordeiro et al. (2009), Hoppe Filho et al. (2017a; b), but not in an expressive and singular way but associated with the filler effect.

Figure 21 shows the chemically combined water content (2C) and portlandite content in the pastes studied at 91 days of hydration, obtained by thermogravimetry. Both are related to Portland cement hydration and are influenced by the occurrence of pozzolanic reaction in cement paste.

The increase in the AQC content and the reduction of the portlandite content of the series with sugarcane bagasse ash in relation to the reference series indicate that the additions showed a higher amount of C-S-H and that this occurred due to the consumption of portlandite typical of the pozzolanic reaction. This corroborates the data obtained by X-ray diffraction which being an indicator of the pozzolanic capacity of the two sugarcane bagasse ash samples from this study.

3.1.2 Determination of Pozzolanlicity by the Modified Chapelle Method

According to research developed by Hoppe Filho et al. (2017) using the modified Chapelle method and according to NBR 15895 (ABNT 2010), the test evaluates the kinetics of pozzolanic reaction in the pozzolan/lime system. At the beginning of the test and during the chemical reactions in the samples, the vitreous fraction of the samples are gradually solubilized by the actions of hydroxyl ions, which react with calcium ions to the precipitations of hydrates (Hoppe Filho 2008). However, according to the results presented in Fig. 22, there is an indication that the SCBA-A has a value lower than the minimum pozzolanic activity index suggested by Raverdy et al. (1980). On the other hand, the SCBA-B ash presents the results above the minimum value of pozzolanic activity index, but very close to this minimum value. Thus, it is possible to conclude that the increase in pozzolanic activity is associated with the amount of amorphous material taking into account the other samples

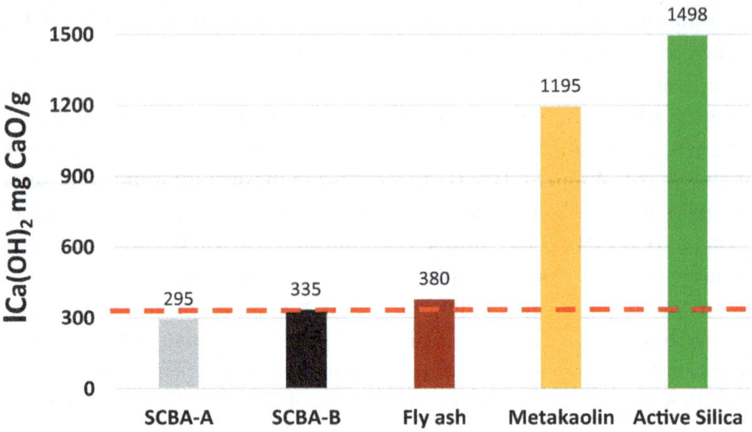

Fig. 22 Chapelle modified test for pozolanicity evaluation

containing a high amount of amorphous material, and the specific surface area of the samples, in addition to the change from quartz-α to quartz-β.

It is noteworthy that NBR 15895 (ABNT 2010) does not define the minimum value for $ICa(OH)_2$ through chapelle's test, however, the French Standard NF P18-513:2010 brings the minimum pozzolanic activity value of 700 mg of $Ca(OH)_2$/g pozzolan, which indicates that the samples SCBA-A, SCBA-B and FA do not fit as pozzolanic material for this test and as suggested by Raverdy et al. (1980) as a minimum value 330 mg of $Ca(OH)_2$/g, which the only sample that would not fit would be the SCBA-A. There is a possibility that values below the minimum established by Raverdy et al. (1980) in the SCBA-A samples may be related to their respective calcination temperatures, due to the direct influence on pozzolanic reactions when these calcination temperatures are different from 600 °C according to Ribeiro and Morelli (2014).

3.2 *Indirect Method of Evaluation of Pozzolanicity*

The results are presented through indirect methods for pozzolanicity analysis.

3.2.1 Electrical Conductivity Analysis

Taking into account the ability of pozzolan to fix calcium hydroxide, thus indicating a loss of electrical conductivity, it is possible to evaluate the pozzolanic activity of a material (Payá 2001). For the analysis of the measurement of the electrical conductivity of a system composed of an unsaturated solution of calcium hydroxide

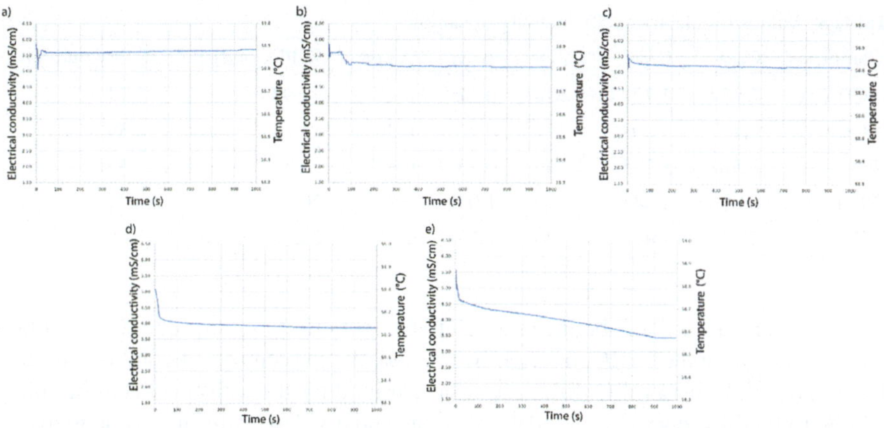

Fig. 23 Electrical conductivity of: **a** SCBA-A ash; **b** SCBA-B ash; **c** Fly ash; **d** Metakaolin and **e** Active Silica

and a fraction of the material evaluated in a time interval from 0 to 1000 s., all materials, procedures and parameters followed those described in 2.4.1

The results of this data collection were the curves of variation of electrical conductivity of sugarcane bagasse ash (SCBA-A and SCBA-B), fly ash (AF), metakaolin (MET) and active silica (SIL), according to Fig. 23. It can be verified that, with the exception of flywheel ash (Fig. 23c), there was a drop in electrical conductivity over time after the insertion of these materials, indicating a pozzolanic potential for all samples. Regarding the fly ash considered as the exception of the samples, Payá (2001) observed the behaviour for measurements up to 1000 s, where there was virtually no conductivity for this sample, but the conductivity according to Payá (2001) studies will occur for longer times (10000). Regarding metakaolin and active silica (Fig. 23d, e), it is observed that conductivity is sharply reduced in the first 20 s of the test and continues to fall slower after this period. On the other hand, all samples show an initial drop in the first twenty seconds.

Taking into account the methodology developed by Payá et al. (2001) the authors use the relative loss of conductivity in the times of 100, 1000 and 10000 s for analysis and evaluations of the pozzolanic activity of the materials.

It was observed that for the samples used in this research that the test up to 1000 s was sufficient and noticeable with regard to a loss of conductivity, attributing such loss to the hypothesis of pozzolanic reactions. According to the method developed by Payá et al. (2001), taking into account the sufficient test time for perception of electrical conductivity, also observed by Basto (2018) and Silva (2016) as the time required for analysis of the samples with a test time interval of 1000s.

Table 8 presents the values related to conductivity immediately before the samples under study (C_{inic}) are inserted, and all conductivity loss data are recorded after the course of 100 and 1000 s of the material inserted in the solution (C100) and (C1000) with variation between the initial and final conductivity for each time 100

Table 8 Values for electrical conductivity of samples

Samples	C_{inic}	C_{100}	C_{1000}	Δ_{100}	Δ_{1000}	% $PRCond_{100}$	% $PRCond_{1000}$
SCBA-A	5.834	5.722	5.674	0.11	0.16	1.92	2.74
SCBA-B	5.427	5.203	5.124	0.22	0.30	4.13	5.59
FA	5.563	5.235	5.145	0.33	0.42	5.89	7.51
MET	5.075	4.034	3.859	1.04	1.22	20.51	23.97
SIL	5.545	4.408	3.443	1.14	2.10	20.50	37.88

and 1000. From the data obtained the relative conductivity losses for 100 and 1000 s (%$PRCond_{100}$) and (%$PRCond_{1000}$). At the end, the percentage of conductivity loss for a given time interval (%PR_{Cond})$_t$ is determined by the ratio between the initial conductivity difference (C_{inic}) and the absolute end divided by the initial conductivity. This percentage will depend on the reaction temperature and the ratio of water, calcium hydroxide and pozzolan contained in the solution (Payá 2001).

In view of the results, it was evidenced that active silica (Fig. 23e) consumes calcium hydroxide extremely quickly, followed by its direct metakaolin competitor (Fig. 23d). It was observed that active silica and metakaolin obtained a much higher loss of conductivity than the other samples in the two times analysed in 100 s and 1000 s as expected, because it is a material considered as super-pozolans well studied.

Given the results for active silica and metakaolin, there is the possibility of the result of the proportions of calcium/pozzolan hydroxide used in the solution influencing the results, also observed by Basto (2018) since, due to the high pozzolanic performance of active silica and metakaolin, the amount of $Ca(OH)_2$ available in the solution would have been quickly fixed by the samples at the initial moments of the test.

According to Payá et al. (2001), when there is a significant loss in electrical conductivity in just a few seconds, that is, in the initial seconds due to a large amount of pozzolan in relation to calcium hydroxide there would be no significant variation in conductivity loss over time. Therefore, it is necessary to study the influence of the proportion of calcium hydroxide/metakaolin to define the amount of ideal pozzolan. According to the research and classification proposal of pozzolanic materials carried out by Lúxan et al. (1989), as expected, the samples of active silica and metakaolin are classified as good pozzolan, with their results found in Table 8 of Δ_{1000} greater than or equal to 1.22 and their results for the loss of conductivity of %$PRCond_{1000}$ 37.91% and 23.96%, respectively.

The results found for the fly ash samples (Fig. 23c) corroborate the results found by Payá (2001), because the researcher found results quite similar to those found in this research. It was also verified according to Payá (2002) that for the fly ash the results up to 1000 s did not significantly express the entire pozzolanic potential of the fly ash, being necessary to extrapolate to times greater than or equal to 10000 s, as suggested by the researcher.

Observing the results presented in Table 8 by the ashes of sugarcane bagasse (SCBA-A and SCBA-B) in Fig. 23a, b respectively and verifying the behavior of electrical conductivity in relation to time, it was identified that for the first 20 s of the test there was a loss in conductivity and that after this time interval the samples showed little variation, corroborating the results obtained by Payá (2002).

These results were also observed by Basto (2018) in his analysis with sewage sludge ash. These initial losses in the electrical conductivity of sugarcane bagasse ash are presented in several samples of sugarcane bagasse ash studied by researchers Janjaturaphan and Wansom (2010), when the tendency of electrical loss in sugarcane bagasse ash samples is noticeable in the first 20 s of the analyses.

The visual analysis of these curves showed that, although Payá et al. (2001) presented their results for the test times of 100, 1000 and 10000 s, in the case of sugarcane bagasse ash, the time of 20 s is also worth mentioning. Therefore, the results of the pozzolanic activity test by electrical conductivity was sensitive to the variations in the ash of sugarcane bagasse in the first 20s.

3.2.2 Portland Cement Performance Index at 28 Days

The samples of sugarcane bagasse ash were evaluated according to NBR 5752 (ABNT NBR 2014), aiming to evaluate their performance index with Portland cement. Tables 9 and 10 and Fig. 24 show the results of axial compressive strength at 28 days of reference mortars without (SCBA) and made with SCBA-A and SCBA-B, materials evaluated in this research. The reference samples (without BCC) had mean mechanical strength of 31.92 MPa and the samples with SCBA-A and SCBA-B presented average mechanical resistances equal to 29.69 MPa and 31.21 MPa, respectively. Bahurudeen and Santhanam (2015), Santos (2017) found higher values of mechanical resistance for cementic matrices containing sugarcane bagasse ash, but the values found are close in the cited authors.

Table 9 shows the results of axial compressive strength of mortars made according to NBR 5752 (ABNT NBR 2014) and Portland cement performance index at 28 days.

Table 9 Results of compressive strength tests

CP'S	Ref.	SCBA-A	SCBA-B	MET	SIL	FA
1	33.30	31.85	30.70	62.31	71.48	60.36
2	30.70	30.20	32.15	60.57	72.01	61.38
3	31.30	29.50	30.80	63.70	70.98	61.13
4	31.60	28.70	30.90	62.65	71.95	60.75
5	32.20	29.80	31.23	62.34	72.44	60.45
6	32.40	28.08	31.45	62.22	71.52	61.37
Average	**31.92**	**29.69**	**31.21**	**62.30**	**71.73**	**60.91**
SD (Standard Deviation)	0.92	1.31	0.54	1.01	0.51	0.45
CV (Coeff. Variation)	2.87%	4.40%	1.73%	1.62%	0.71%	0.74%

Table 10 Summary of performance index at 28 days

Mixtures	Compressive strengths at 28 days in MPa	Portland cement performance index
Reference	31.92	1.00
SCBA-A	29.69	0.93
SCBA-B	31.21	0.98
FA	60.91	1.91
MET	62.30	1.95
SIL	71.73	2.25

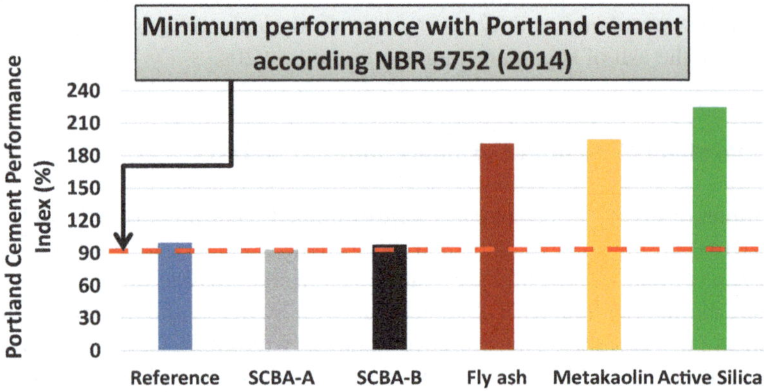

Fig. 24 Portland Cement Performance Index

In Table 10, abstracts are presented with the percentage of the performance index and Fig. 24 clearly portrays these percentages with the minimum performance required according to NBR 5752 (ABNT NBR 2014). These indices of Pozzolanic Activity with cement (IAP) of mortars containing SCBA-A and SCBA-B are presented.

According to the results obtained, it is observed that the ashes of sugarcane bagasse (SCBA-A and SCBA-B) presented satisfactory results, that is, above the limit suggested by the standard. According to these results, it can be affirmed that the ashes studied presented satisfactory performance indexes, with a small difference between the two ashes ranging around 5% (Figs. 25 and 26).

The results of mechanical resistance corroborate the x-ray and thermogravimetric diffraction analyses performed and presented in the previous sub-sections, indicating that these ashes present pozzolanic reactivity. In addition, it was possible to observe in this method that there was no significant influence between the two ashes studied (SCBA-A and SCBA-B), comparing the two ashes to each other, but there were differences. Corroborating the results found by the authors Bahurudeen and Santhanam (2015), who found significant influence on the pozzolanicity of the ashes studied

Fig. 25 **a** SCBA-A ash—SEM MAG (2.50kx); **b** SCBA-B ash—SEM MAG (500x); **c** Fly ash—SEM MAG (5.00kx); **d** Metakaolin—SEM MAG (5.00kx) and **e** Active silica—SEM MAG (2.50kx)

due to temperature variation in calcination, were also observed in this research in relation to the samples SCBA-A and SCBA-B.

It is important to highlight that pozzolanic activity alters the microstructure of the cement matrix, through the chemical reactions between portlandite and pozzolan, which is a compound generated by stable secondary hydrated products. Consequently, this phenomenon causes the refinement of the pores, increasing the tortuosity of the capillary network, besides densifying the transition zone, increasing the mechanical resistance of the cement matrix. This phenomenon is attributed to the refinement of the microstructure due to the formation of crystals of secondary hydrated calcium silicates (C-S-H). In addition to the chemical effect, pozzolanic materials can increase the mechanical resistance of cementmarices through the filler effect (Hoppe Filho 2008).

3.3 Scanning Electron Microscopy Analysis

The results are presented using the scanning electron microscope.

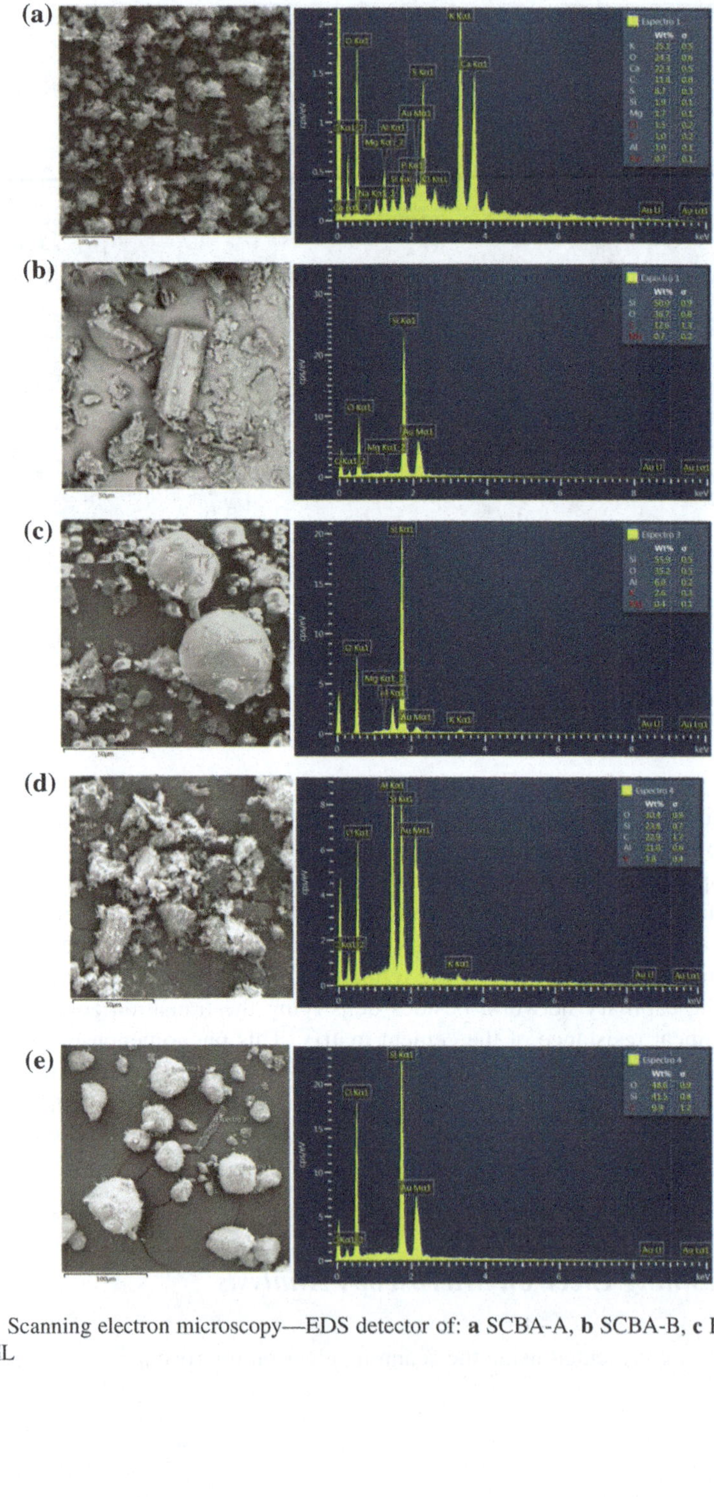

Fig. 26 Scanning electron microscopy—EDS detector of: **a** SCBA-A, **b** SCBA-B, **c** FA, **d** MET and **e** SIL

3.3.1 Scanning Electron Microscopy Analysis—SE Detector

According to Mehta and Monteiro (2014) the C-S-H phase comprises approximately 50% to 60% of the volume of solids in a hydrated paste, being undoubtedly the most important and determining phase of its properties. At a nanometric or atomic level, hydrated calcium silicate (C-S-H) is formed by layers with SiO4 chains with possible changes, i.e., replacing silicon with aluminum.

Mehta and Monteiro (2014) showed that portlandite (CH) presents a characteristic shape of hexagonal prismatic crystals, constituting about 20% to 25% of the volume of solids in hydrated cement paste. The formations of these crystals are affected by the availability of space, by the temperature at which cement hydration is being performed, among other factors that act directly in this formation. On the other hand, portlandite contributes in a little expressive way to the mechanical strength of concrete, mainly because its surface area is considerably very low.

3.3.2 Scanning Electron Microscopy Analysis—EDS Detector

3.4 Absorption, Void Index and Specific Mass

3.4.1 Immersion Water Absorption Test, A(%)

According to NBR 9778 (ABNT NBR 2009) and its respective international standard ASTM C29/C29M (C29/C29M 2017), it is possible to determine and analyse through this test of the behaviour of the specimens in relation to water absorption by immersion, making a comparison analysis between the samples with the addition of 15% of the materials and the reference samples. Figure 27 shows the results of the samples in that test.

As can be observed, it can be noted that the reference samples obtained the highest average in terms of absorption as expected. As also expected, the samples of active silica obtained the lowest absorption by immersion, corroborating the researches of Cordeiro (2006). The samples using 15% of sugarcane bagasse ash addition obtained acceptable results throughout the trials and in their final result. These results are very coherent in relation to the research developed by Sperbe (2003), Cordeiro (2006), when the researchers achieved very similar results. This test provided us with information on how much each sample absorbed from water being immersed for 72 continuous hours, as well as the void index and specific mass of each specimen.

3.4.2 Determination of Void Indexes, Iv(%)

According to NBR 9778 (ABNT NBR 2009) and its respective international standard ASTM C29/C29M (C29/C29M 2017), it was possible to determine and analyse through this test the behaviour of the specimens in relation to their void index present

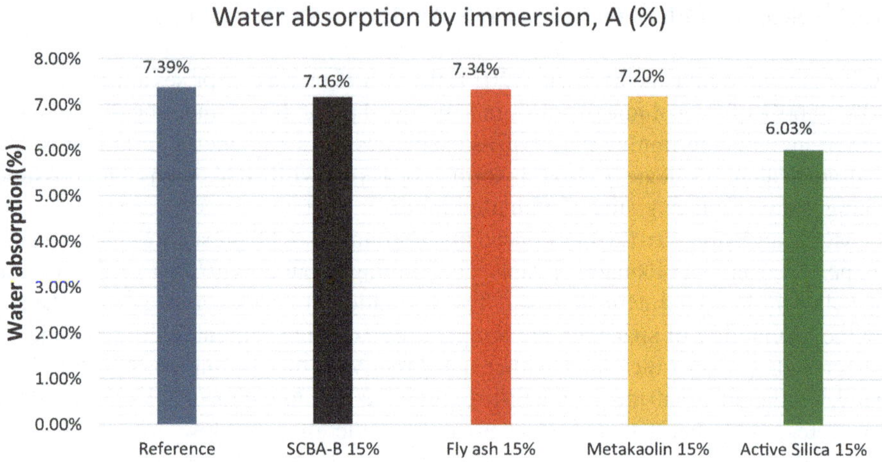

Fig. 27 Water absorption by immersion, A(%)

in the samples, making a comparison analysis of the samples with the addition of 15% of the materials studied with the reference samples. Figure 28 shows the results of the samples in that test.

As can be observed, it can be observed that the reference samples obtained the highest average in terms of void index as expected, as well as the active silica samples obtained the lowest void index, being totally cohesing this result in relation to the result obtained by Cordeiro et al. (2009), corroborating by Santos (Santos 2017) research. The samples of active silica, in addition to the chemical reactions with pore

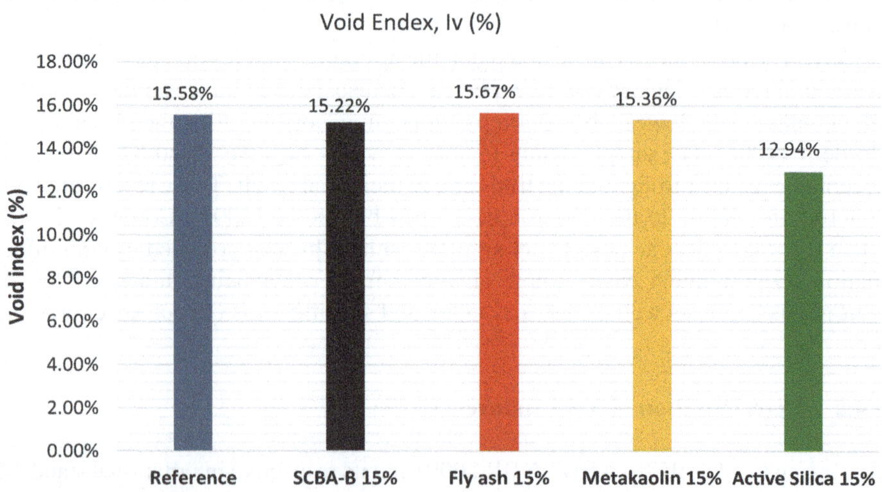

Fig. 28 Void index I_V(%)

refinement, also have a physical effect, i.e., filler effect confirming the results. The samples using 15% of sugarcane bagasse ash added 15% obtained a very satisfactory result in the test and in its final result, because it was observed that the SCBA presented a lower average void index than the samples with the addition of 15% fly ash. These results are extremely consistent with the researches developed by Sperbe (2003), Cordeiro (2006), in which the researchers obtained very similar results.

3.4.3 Determination of the Specific Mass

According to NBR 9778 (ABNT NBR 2009) and its respective international standard ASTM C29/C29M (C29/C29M 2017), it was possible to determine and analyse through this test the behaviour of the specimens in relation to the specific dry mass present in the samples and to make a comparison analysis of the samples with the addition of 15% of the materials studied with the reference samples. Figure 29 shows the results of the samples in that test.

As observed, it is notorious that the reference samples obtained the lowest mean of the specific dry mass as expected, as well as the active silica samples obtained the highest mean specific dry mass, being totally consistent with the results of Cordeiro (2006) and corroborating the researches developed by Santos (2017). The samples using 15% of the addition of sugarcane bagasse ash obtained a very satisfactory result in the test and in its final result, because it was observed that the SCBA presented specific dry mass very close to the values of the other samples, that is, the samples with the addition of 15% fly ash and 15% metakaolin. These results are very coherent in relation to the researches developed by Sperbe (2003), Cordeiro (2006) because the researchers obtained extremely similar results.

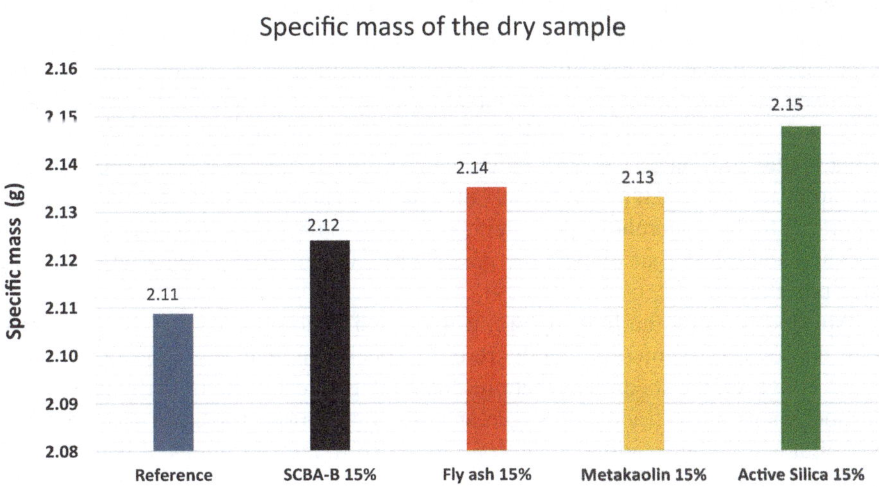

Fig. 29 Specific mass of the dry samples

The specific dry masses obtained through the test showed variations in relation to the samples studied. This is due to the different water losses to which they are submitted once the hardening and hardening process in cement hydration has begun. The values obtained in the tests of void indices corroborate the idea that concrete with smaller specific masses have a higher void index in its structure, besides presenting higher values of water absorption. The research was also the object of durability when evaluating capillary absorption rates, with the main objective of verifying the properties involved and their reflex evidenced in their capillary absorption.

In the specific case of immersion absorption, the relationship of greatest influence observed is associated with specific dry mass, but in addition, the water/cement ratio is much more influential than in the absorption tests, void index and specific dry mass, according to the research developed by Berenguer et al. (2018).

Regarding the absorption phenomena, the evaluated concretes showed an increase in immersion absorption with the decrease of their specific masses. On the other hand, there is the knowledge that the permeability of concrete is a crucial factor for their durability; the more permeable the concrete, the more susceptible to the deleterious actions of aggressive agents present in the environmental atmosphere. Therefore, special attention should be given when using concrete that have high void indexes and high capillarity, according to what was observed in the study.

3.4.4 Resistance to Accelerated Carbonation

Mineral additions generally improve the properties of concrete, especially pozzolanic additions, including properties related to its durability. On the other hand, some researchers report that carbonation can be negatively affected, bringing some unfavorable results. The studies developed by Khunthongkeaw et al. (2006), Rukzon and Chindaprasirt (2010), Younsi et al. (2013), show that the use of fly ash, a pozzolan of medium reactivity, considerably increases the carbonation front speed. This result is possible when using pozzolan in partial replacement of Portland cement, being explained by the reduction of the alkaline reserve of concrete. The research of Turk et al. (2013) has been evidenced that the use of super-pozolans i.e., active silica, metakaolin and rice husk silica, when using in partial cement replacement, may improve the resistance to the advancement of carbonation in concrete, because according to the researchers, pozzolanic reactions refine pores of such order through the chemical effect that hinders the passage of CO_2 through pores. The results are shown in Fig. 30.

After performing the first rupture, it was possible to observe that the specimens with the lowest carbonation depth were the specimens with the addition of active silica. However, the specimens that presented the greatest depth of carbonation were the specimens of the trace with sugarcane ash, in these specimens, at some points it was obtained higher readings than those recorded in the specimens of the reference trace.

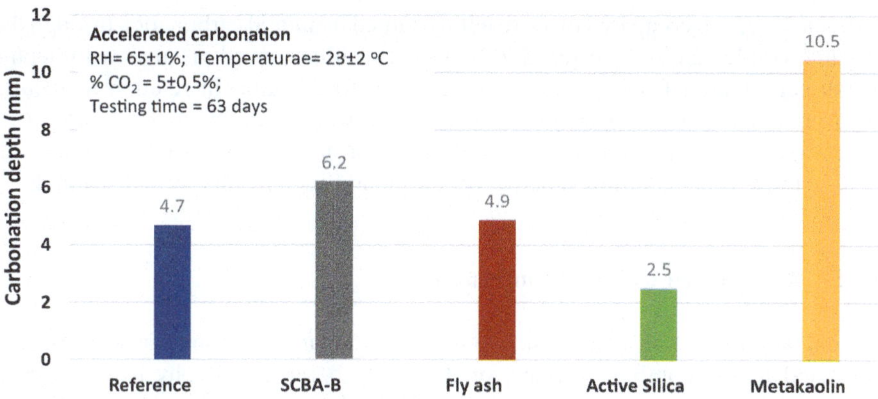

Fig. 30 Accelerated carbonation—63 days of testing

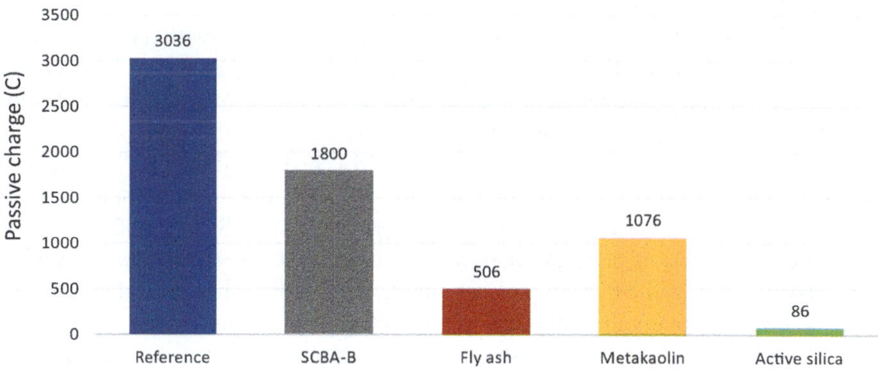

Fig. 31 Passing charge of accelerated penetration of chloride ions

For the specimens with 15% of the addition of fly ash, carbonation advances were observed in some points that were higher than the reference samples. In the second rupture of the samples, the trends of the carbonation front advance obtained a continuity, and it was observed that the samples with active silica obtained their values of the advance of the carbonation front at 63 days at 2.5 mm, for the samples containing fly ash obtained an advance of 4.9 mm, followed by the ashes of sugarcane bagasse with 6.2 mm. On the other hand, in relation to the advancement of the carbonation front for the reference samples obtained values of 4.7 mm, standing out positively in relation to the fly ash and ash of sugarcane bagasse, and these results are fully expected according to researchers (Khunthongkeaw et al. 2006; Rukzon and Chindaprasirt 2010; Younsi et al. 2013).

Therefore, the samples containing the ashes of sugarcane bagasse, taking into account the attack by aggressive CO2 agents, are vulnerable to such attack, providing progress in the carbonation front. However, regarding the results of the accelerated carbonation front, compared to the results obtained in natural carbonation through the research conducted by Dunster (2000), all the results showed promise with depths of the carbonation front ranging from 2.5 mm to 6.2 mm. In relation to natural exposure one can hypothesized that the additions were favorable to the durability of the concrete. It is also important to highlight that in the tests for the samples with 15% of metakaolin addition, an error occurred, and such results were discarded.

3.4.5 Accelerated Chloride Ion Penetration

The degradation of reinforced concrete structures occurs due to some aspects, mainly attributed to the combined effects of different chemical and physical processes involved. Reinforcement corrosion is invariably the main cause of degradation of reinforced concrete structures (Mehta and Monteiro 2014). According to Sitter (1984) this type of pathological manifestation is responsible for high financial expenses for the structural recovery of buildings.

Verifying the values of void index and capillary absorption, with the additions of 15% of the materials studied in this research (see Figs. 27 and 28), it is observed that they did not compromise the accelerated penetration values of chloride ions, being observed falls in the passing loads of these samples.

According to the criteria established in ASTM C1202 (ASTM 1202), the values obtained with the test are described in Fig. 28. The reference samples (REF), according to the criteria established in ASTM C1202 (ASTM 1202), obtained moderate classification as an evaluation criterion, the SCBA-B samples obtained a low classification, while the samples containing 15% of fly ash (FA) and metakaolin (MET) addition obtained a very low classification in relation to the samples with 15% of active silica addition that stood out with the insignificant classification and with passing electrical load of 86 °C (Fig. 31).

In continuity with the analyses and their respective tests and results, after performing the accelerated penetration tests of chloride ions according to ASTM C1202 (ASTM 1202), the colorimetric indicators by silver nitrate sprinkling were performed (see Figs. 32 and 33). Silver nitrate sprinkling was performed in order to verify the thickness of the layer contaminated by chloride ions in the samples. For this purpose, it is necessary to centrally break the samples, using the apparatus for the compressive strength test.

After the accelerated chloride penetration test following ASTM C1202 (ASTM 1202) in the samples at 365 days of age and performed by Sprinklings of $AgNO_3$, the results were obtained, as highlighted in Fig. 33. The results presented include an increase in the depth of the chloride-contaminated layer for the reference samples, as already identified in Fig. 31 through passing loads, corroborating this aspect. Another issue identified was the reduction of the depth of penetration with the use of 15% of the mineral additions, as can be seen in Fig. 33, i.e., all samples with 15% addition obtained values lower than 10 mm and the samples with 15% of active silica again stand out positively in relation to the others.

Fig. 32 Test with colorimetric indicators by silver nitrate sprinkling

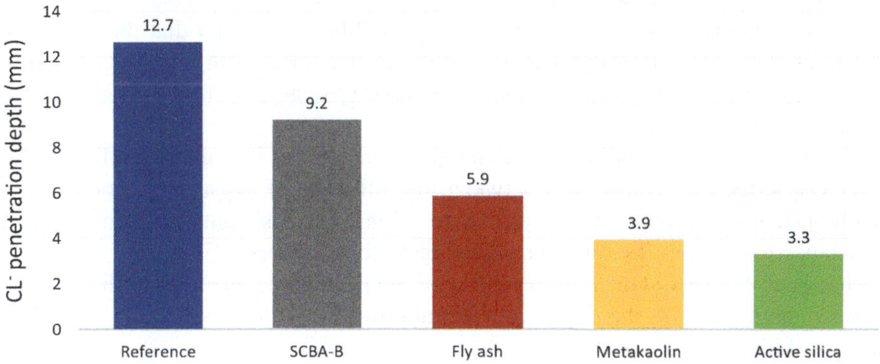

Fig. 33 Depth of penetration of chloride ions

It is also noted that when analyzing chloride ion penetration (Fig. 33) and capillary absorption the samples behaved in a natural tendency for the correlation between the two tests, where the two extremes, with the reference samples standing out negatively for both tests and the samples with active silica standing out positively. And the other samples were very close in terms of values, with emphasis also on the samples with 15% addition of metakaolin. Highlighting the resistance to chloride ions penetration for samples with 15% mineral addition, raising the hypothesis of the use of 15% addition of sugarcane bagasse ash as a promising material in terms of durability to chloride ion attack.

In view of the accelerated penetration results with the passing loads in the samples and the depth of penetration of chloride ions, through the test of colorimetric indicators by silver nitrate sprinkling, there is a positive trend between the two analyses (see Fig. 34).

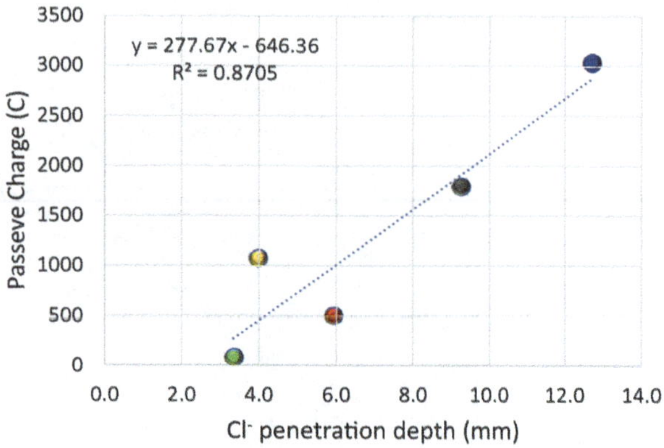

Fig. 34 Passing charge versus chloride ions penetration depth

In the case of the classic migration test according to ASTM C1202 (ASTM 1202), the results indicate the existence of a considerable correlation ($R^2 = 0.8705$) between the values of passing load and chloride penetration depth by the visual colorimetric method. This indicates that, although it is a qualitative test and that the results do not refer exclusively to the transport of chloride ions, according to Medeiros et al. (2009) the response variable of this test is directly related to the depth of chloride penetration in concrete.

The results confirm the results obtained by researchers Rodolpho and Isa (1999), which describe the relationship between the influence of resistance to penetration to chloride ions on the durability of concrete constructions, being fundamental such resistance to this harmful and aggressive agent to structures.

It was also observed the relationship between pozzolanic reactions with the modified Chapelle test and the depth of penetration of chloride ions, observed as these relationships according to Fig. 35.

For the case of the classic migration test according to ASTM C1202 (ASTM 1202), the results indicate the existence of a considerable correlation ($R^2 = 0.7571$)

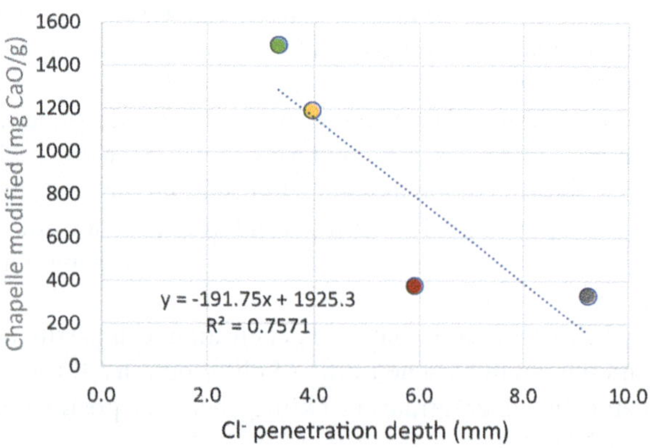

Fig. 35 Chapelle modified versus chloride ions penetration depth

between the values of the modified Chapelle test and the depth of chloride penetration by the visual colorimetric method. This indicates that, although it is a qualitative test, there is a strong trend between the two techniques, which are directly related to the durability of concrete structures observed by researchers Swamy et al. (1998) and Vilarino et al. (1995). Results above 1000 mg CaO/g in the modified Chapelle indicates a depth of penetration of chloride ions smaller than or equal to a maximum of 4.0 mm. Sendo considerados valores insignificantes em relação à durabilidade das estruturas, de acordo com Swamy et al. (Swamy et al. 1998), para os demais valores para as amostras de cinza volante com uma profundidade de íons cloretos de 5.9 mm e para as amostras de cinza de cana-de-açúcar de 9.2 mm, são consideradas de 9.2 mm, são consideradas de 9.2 mm, são consideradas de com os pesquisadores Swamy et al. (1998), Vilarino et al. (1995) com baixa penetração, sendo favorável à durabilidade das estruturas de concreto.

3.5 Analysis of the Results Obtained

Taking into account the results obtained through various techniques, using various methods and procedures, several variables are created that hinder correlations between accelerated and non-accelerated techniques. It is also important to highlight that the difficulty of correlating the methods proposed for the evaluation of pozzolanic reactions are enhanced due to the high complexity of the phenomenon.

In the course of the research it was observed that these methods are performed in different experimental systems and conditions. In addition, it was possible to observe that some tests are performed in a long period of time (non-accelerated methods), such as the pozzolanic performance index with Portland cement, which are evaluated at 28 days; and other techniques in a short period of time (accelerated methods), for example electrical conductivity that counts evaluations in 100 s and 1000 s. Considering these limitations, we chose to correlate these methods separately only if these correlations or trends are possible.

3.5.1 Materials Characterization

Considering that different materials with different properties were used, it was possible to clearly observe, using the scanning electron microscopy technique through the SE detector, the morphology of the samples, being observed how much they differ from each other. It was also observed in relation to the particle size distribution that some results were not expected according to the summary of Table 11, but totally justified when associated with the two techniques, and it is noticeable that some samples clustered (using the technique by scanning electron microscopy such agglomeration was detected). When analysing the specific surface area, as before, some samples presented surprising results according to the summary of Table 11, but when using other techniques (scanning electron microscopy), the results become

Table 11 Summary of the materials characterization

Properties	Cement	SCBA-A	SCBA-B	FA	MET	SIL
Specific surface area (m^2/g) (Blaine Electronic)	0.452	0.655	0.654	0.678	0.474	0.718
Specific surface area (m^2/g) (BET)	1.784	5.461	22.849	1.158	16.908	20.341
Particle distribution D_{50} μm	5.90	8.25	10.32	9.76	5.72	63.67
Mass loss % (TG)	6.08	22.22	11.65	1.34	2.63	2.98
Fire loss % (FL)	6.37	19.89	13.64	2.03	3.07	3.09

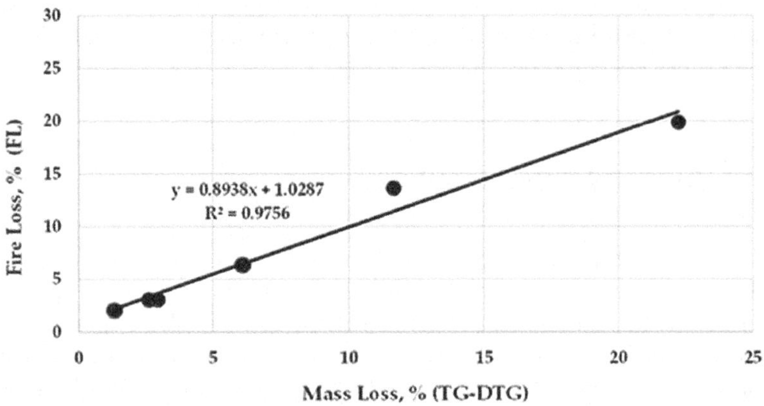

Fig. 36 Fire Loss % (FL) versus Mass Loss % (TG)

justifiable, because some samples presented shapes with cavity, sticks, irregular shapes and etc. Making your specific area larger by percolation of gas in the B.E.T. technique.

In relation to thermoanalysis, the correlation between the two techniques is observed, that is, using fire loss and thermogravimetry, where there is a correction ($R^2 = 0.9756$) in their results according to Fig. 36, in view of the results observed at the approximation of the techniques. Subsequently, the chemical elements of each sample are analysed, being observed what the standard NBR 12653 (ABNT 1265) recommends, in terms of oxides and to support the subsequent x-ray diffraction technique, given the large amount of silicon oxide present in the sample it was possible to quantify the amount of amorphous material through Rietveld refinement.

3.5.2 Evaluation of Direct Method Pozzolanicity

In view of the results with the characterizations of the materials, the analyses of the pozonalicities of the samples were made, that is, direct and indirect analyses. In the direct analyses of the pozolancity of the analysed materials, the thermoanalysis

Table 12 Summary of chemically combined water and portlandite [AQC and $Ca(OH)_2$]

Cement Paste	Cure	$AQC_{(\%)}$	$Ca(OH)_2$
Cement CP V—ARI	**28 days**	10.69	17.28
SCBA—A		10.28	25.00
SCBA—B		10.73	21.42
Cement CP V—ARI	**63 days**	10.70	19.48
SCBA—A		13.11	25.29
SCBA—B		10.96	18.37
Cement CP V—ARI	**91 days**	10.83	24.86
SCBA—A		13.11	23.54
SCBA—B		11.67	17.66

Table 13 Summary modified Chapelle method

	SCBA-A	SCBA-B	FA	MET	SIL
ICa(OH)2 mg CaO/g modified Chapelle	295	335	380	1195	1498

technique was sensitive to the perception of portlandite consumption by chemical reactions, being possible to quantify through stochiometric calculations the loss of mass in the thermal event, where the decomposition of portlandite occurs according to the summary of Table 12.

Subsequently, such chemical reactions were also observed using the modified Chapelle method, where all samples with 15% addition obtained performance above the minimum (330 mg/g) expected according to Raverdy et al. (1980), with the exception of SCBA-A samples, falling slightly below according to the summary of Table 13.

3.5.3 Evaluation of Indirect Method Pozzolanicity

In continuity with the research, the analyses were performed through indirect methods with the analysis by electrical conductivity and performance index with Portland cement, in which the electrical conductivity was observed that the analyzed materials corroborated the results found in the researches of Lúxan et al. (1989), Payá (2001).

The loss of conductivity for the samples was noticeable according to Table 14, however, the SCBA-A and SCBA-B samples do not fit the classification proposed by Lúxan et al. (1989), on the other hand, it is important to highlight that the researcher evaluated in his research natural pozzolanic materials and that Payá (2001) showed that the time of analysis for fly ash should be much longer than 120 s (the one proposed by Lúxan et al. 1989). In continuity, the researches Payá et al. (2001) and Payá et al.

Table 14 Summary of electrical conductivity and performance index

Samples	Electrical conductivity		Performance index with Portland cement (%)
	Δ_{100}	Δ_{1000}	
REF	–	–	100
SCBA-A	0.11	0.16	93
SCBA-B	0.22	0.30	98
FA	0.33	0.42	–
MET	1.04	1.22	–
SIL	1.14	2.10	–

(2001) observed the behaviour using sugarcane bagasse ash and verified that studies with BCC should be deepened to arrive at a more appropriate classification parameter.

Therefore, the classifications proposed by Lúxan et al. (1989) should not be considered in the analyses of sugarcane bagasse ash samples. On the other hand, should be mentioned that there are changes in the baseline of all graphs represented in Fig. 23 for all samples with their respective losses of electrical conductivity, thus raising the hypothesis that the ashes of sugarcane bagasse reacted chemically, causing the losses of electrical conductivity.

Following the analyses, for the Portland cement performance index test, it was observed that the ashes were above the minimum limit of 90% (0.90) allowed by the standard. These results corroborate the modified Chapelle test, in which all samples were above the minimum limits (330 mg/g) established by Raverdy et al. (1980), with the exception of the SCBA-A sample, but with values very close to this minimum established according to the summary of Table 14.

Therefore, from all the results discussed, it can be concluded that there is no evident correlation between the accelerated methods (Modified Chapelle and electrical conductivity) that were used during this research, different from that occurred between the non-accelerated methods, which presented a good correlation.

3.5.4 Evaluation and Analysis of Concrete Durability

For the immersion absorption tests, the samples containing 15% of addition obtained good results in relation to the reference samples, thus showing that the use of 15% of sugarcane bagasse ash, in relation to capillary absorption and immersion, performs a refinement of capillary pores according to Table 15.

Regarding the void indexes, the samples were also better than the reference ones, these results become important from the point of view of durability, because they positively inhibit the entry of aggressive agents in reinforced concrete structures according to Table 15.

Considering the results obtained with the accelerated carbonation test, we observed that all additions are favourable to the durability of concrete structures,

Table 15 Durability analysis summary

Samples	Absorption by immersion (%)	Void indexes (%)	Carbonation depth (mm)	Passing charge (C)	Chloride depth (mm)
REF	7.39	15.58	4.7	3036	12.7
SCBA-B	7.16	15.22	6.2	1800	9.2
FA	7.34	15.67	4.9	506	5.9
MET	7.20	15.36	–	1076	3.9
SIL	6.03	12.94	2.5	86	3.3

although the ash (FA and SCBA-B) presented somewhat unfavourable results in relation to reference samples from the point of view of accelerated carbonation according to Table 15. However, fully coherent and predicted according to the results obtained by Turk et al. (2013), due to the consumption of the alkaline reserve. Its absorptions by immersion were favourable, raising the hypothesis of capillary pore refinement.

According to the results obtained with the accelerated penetration of chloride ions, the samples with the addition of 15% of the studied materials were totally favourable to the durability of the structures, being observed the reduction of penetration for all samples with the addition in relation to the reference samples, corroborating the results presented by the researchers Swamy et al. (1998), Vilarino et al. (1995).

4 Conclusions

In view of all the analyses and the results expressed, it is possible to reach the following conclusions:

It is evident that mineral additions play significant gains in the basic properties of mortars and concrete, improving mechanical properties for example, refining their capillary pores, among others. In view of the results of the analyses for pozzolanic determination of the samples of sugarcane bagasse samples SCBA-A and SCBA-B, it was observed that for some trials the SCBA-A was considered as pozzolanic material, but in other analyses considered with low reactivity, not reaching the minimum expected.

It is important to highlight that the SCBA-A and SCBA-B samples did not have any type of processing, besides being dried and sifted, in the 0.075 mm opening sieve; therefore, given the results obtained, the ash of sugarcane bagasse SCBA-A can be considered as a material that has low pozzolanicity. It is possible to use it as pozzolanic material, but with some improvements, such as calcination and grinding to increase in its specific surface area, increasing its reactivity. Regarding the ash samples of sugarcane bagasse SCBA-B, they were considered as pozzolanic material of medium reactivity, because in the analyses and tests performed they obtained results higher or close to the minimum of expected performance and being susceptible to possible uses as mineral addition.

Therefore, from all the results discussed, it can be concluded that there is no evident correlation between the accelerated methods (Modified Chapelle and electrical conductivity) that were used during this research. However, in relation to the non-accelerated methods, some techniques presented a good correlation.

As the SCBA-B obtained satisfactory results, it becomes possible to use it maintaining the same conditions of the research, with proportions of 15% of addition as shown throughout the research.

It can be concluded, in terms of durability, that the samples containing 15% of the ash of sugarcane bagasse showed good results, but it is important to highlight that the initial analyses of durability of mortars and concrete, despite obtaining satisfactory results in some aspects, require further research on the durability of the structures when added the ashes of sugarcane bagasse, because several points in relation to durability need to be explored and evaluated with technical depths.

It can finally be concluded that the phenomena that occur during the pozzolanic reaction are complex, which makes it difficult to determine the reactivity of a given material, being possible only to identify general trends of its behaviour.

Another important point is the sustainability that the use of sugarcane bagasse ash provides. This material is often disposed of in dumps and has no use. When its incorporation into cement matrices, a new reuse arises, given that bagasse is not the raw material of sugar production and is already used to generate energy, having as tailings ash, which in turn will be reused in the production of concrete and mortars.

References

ABNT NBR 12653:2014 Pozzolanic materials—Requirements Available online: http://www.consorciobradesco.com.br/. Accessed Jun 17, 2020

ABNT NBR 15895:2010 Pozzolanic materials—Determination of calcium hydroxide fi xed—Modified Chapelle´s method Available online: https://www.abntcatalogo.com.br/norma.aspx?ID=80644. Accessed Jun 17, 2020

ABNT NBR 16372:2015 Portland cement and other powdered materials—Determination of fineness by the air permeability method (Blaine method) Available online: https://www.abntcatalogo.com.br/norma.aspx?ID=332676. Aaccessed Jun 17, 2020

ABNT NBR 16605:2017 Portland cement and other powdered material—Determination of the specific gravity Available online: https://www.abntcatalogo.com.br/norma.aspx?ID=372002. Aaccessed Jun 17, 2020

ABNT NBR 9778:2009 Hardened mortar and concrete—Determination of absorption, voids and specific gravity Available online: https://www.abntcatalogo.com.br/norma.aspx?ID=52163. Accessed Jun 17, 2020

ABNT NBR 9779:2012 Mortar and hardened concrete—Determination of water absorption by capillarity Available online: https://www.abntcatalogo.com.br/norma.aspx?ID=193710. Accessed Jun 17, 2020

ABNT NBR 5752:2014 Pozzolanic materials—Determination of the performance index with Portland cement at 28 days Available online: https://www.abntcatalogo.com.br/norma.aspx?ID=315344. Accessed Jun 17, 2020

ABNT NBR 7214:2015 Standard sand for cement tests—Specification Available online: https://www.abntcatalogo.com.br/norma.aspx?ID=348035. Accessed Jun 17, 2020

ABNT NBR 5739:2018 Concrete—Compression test of cylindrical specimens Available online: https://www.abntcatalogo.com.br/norma.aspx?ID=398444. Accessed Jun 17, 2020
ABNT NBR 7215:2019 Portland cement—Determination of compressive strength of cylindrical test specimens Available online: https://www.abntcatalogo.com.br/norma.aspx?ID=413557. Accessed Jun 17, 2020
ABNT NBR 5738—Concrete—Procedure for molding and curing prov bodies a. Rio de Janeiro, 2015, corrected on 28/06/2016
ABNT NBR 6118—Design of concrete structures—Procediment. Rio de Janeiro, 2014
Araya-Letelier G, Maturana P, Carrasco M, Antico FC, Gómez MS (2019) Mechanical-damage behavior of mortars reinforced with recycled polypropylene fibers. Sustain, 11. https://doi.org/10.3390/su11082200
Arif E, Clark MW, Lake N (2017) Sugar cane bagasse ash from a high-efficiency co-generation boiler as filler in concrete. Constr Build Mater 151:692–703. https://doi.org/10.1016/j.conbuildmat.2017.06.136
ASTM C1202-19 Test Method for Electrical Indication of Concretes Ability to Resist Chloride Ion Penetration Available online: astm.org/Standards/C1202. Accessed Jun 17, 2020
ASTM C29/C29M (2017) Standard Test Method for Bulk Density ("Unit Weight") and Voids in Aggregate. ASTM, West Conshohocken
Azmee NM, Shafiq N (2019) Investigating the impacts of ultra-fine calcium carbonate in high-volume fly ash concrete for structural rehabilitation for sustainable development. Sustain. https://doi.org/10.3390/su11174671
Bahurudeen A, Santhanam M (2015) Influence of different processing methods on the pozzolanic performace of sugar cane bagasse ash. Cement Concrete 56:32–45
Basto PEA (2018) Determinação da pozolanicidade por condutividade elétrica do lodo de estações de tratamento de esgoto (ETE) para adição ao Cimento Portland. Dissertação (mestrado)—Universidade Federal de Pernambuco. Recife
Berenguer RA, Nogueira SFA, Barreto MEC, Lins CS, Lima A (2018) Comparative evaluation of the mechanical behavior of light concrete with incorporated air. Revista Alconpat 8(2):178–193
Carvalho SZ, Vernilli F, Almeida B, Oliveira MD, Silva SN (2018) Reducing environmental impacts: The use of basic oxygen furnace slag in portland cement. J Clean Prod 172:385–390. https://doi.org/10.1016/j.jclepro.2017.10.130
Cascudo O, Carasek H (2014) Durability of concrete: scientific basis for the formulation of durable concretes according to the environment, 1st edn. IBRACON, São Paulo, SP
Chen C, Habert G, Bouzidi Y, Jullien A (2010) Environmental impact of cement production: detail of the different processes and cement plant variability evaluation. J Clean Prod 18:478–485. https://doi.org/10.1016/j.jclepro.2009.12.014
Chen HJ, Shih NH, Wu CH, Lin SK (2019) Effects of the loss on ignition of fly ash on the properties of high-volume fly ash concrete. Sustain, 11. https://doi.org/10.3390/su11092704
Cordeiro. G. C. Utilização de Cinza Ultrafina do Bagaço de Cana de Açúcar e de Casca de Arroz como Aditivos Minerais em Concreto. Tese de Doutorado—Programa de Pós-Graduação Engenharia Civil, Universidade Federal do Rio de Janeiro (2006)
Cordeiro GC, Toledo Filho RD, Tavares LM, Fairbairn EMR (2009) Ultrafine grinding of sugar cane bagasse ash for application as pozzolanic admixture in concrete. Cement Concrete Res 39(2):110–115
Duart MA (2008) Study of Concrete Microstructure with Addition of Residual Rice Husk Ash without Beneficiation. MSc Thesis. Universidade Federal de Santa Maria. Santa Maria, 2008
Dunster AM (2000) Accelerated carbonation testing of concrete. BRE—Information papers of the building research establishment, edição: 20
Dweck J, Cherem da Cunha AL, Pinto CA (2009) Pereira Gonçalves, J.; Büchler, P.M. Thermogravimetry on calcined mass basis—hydrated cement phases and pozzolanic activity quantitative analysis. J Therm Anal Calorim, 97, 85–89. https://doi.org/10.1007/s10973-008-9761-0

Elizondo-Martinez EJ, Tataranni P, Rodriguez-Hernandez J, Castro-Fresno D (2020) Physical and mechanical characterization of sustainable and innovative porous concrete for urban pavements containing metakaolin. Sustain, 12. https://doi.org/10.3390/su12104243

Gursel AP, Ostertag CP (2016) Impact of Singapore's importers on life-cycle assessment of concrete. J Clean Prod 118:140–150. https://doi.org/10.1016/j.jclepro.2016.01.051

Helene PRL, Ribeiro DV (Org.) Corrosion in reinforced concrete structures: Theory, control and methods of analysis.1ª Edição, Rio de Janeiro: Elsevier

Hoppe Filho J (2008) Cement, fly ash and hydrated lime systems: hydration mechanism, microstructure and carbonation of concrete. PhD Thesis, Escola Politécnica da Universidade de São Paulo, São Paulo, Brazil

Hoppe Filho J, Gobbi A, Pereira E, Quarcioni VA, Medeiros MHF (2017a) de Atividade pozolânica de adições minerais para cimento Portland (Parte I): Índice de atividade pozolânica (IAP) com cal, difração de raios-X (DRX), termogravimetria (TG/DTG) e Chapelle modificado. Matéria (Rio Janeiro), 22. https://doi.org/10.1590/s1517-707620170003.0206

Hoppe Filho J,Gobbi A, Pereira E, Tanaka RS, Medeiros MHF (2017b) Atividade pozolânica de adições minerais para cimento Portland (Parte II): Índice de atividade pozolânica com cimento Portland (IAP), difração de raios-X (DRX) e termogravimetria (TG/DTG). Matéria-Rio de Janeiro, v. 22, p. e11873, 2017(b)

Hossain MU, Poon CS, Lo IMC, Cheng JCP (2017) Comparative LCA on using waste materials in the cement industry: A Hong Kong case study. Resour Conserv Recycl 120:199–208. https://doi.org/10.1016/j.resconrec.2016.12.012

Hossain MU, Poon CS, Dong YH, Xuan D (2018) Evaluation of environmental impact distribution methods for supplementary cementitious materials. Renew Sustain Energy Rev 82:597–608. https://doi.org/10.1016/j.rser.2017.09.048

Huang TY, Chiueh PT, Lo SL (2017) Life-cycle environmental and cost impacts of reusing fly ash. Resour Conserv Recycl 123:255–260. https://doi.org/10.1016/j.resconrec.2016.07.001

Janjaturaphan S, Wansom S (2010) Pozzolanic activity of industrial sugar cane Bagasse Ash. Suranaree J Sci Technol 17(4):349–357

Khunthongkeaw J, Tangtermsirikul S, Leelawat T (2006) A study on carbonation depth prediction for fly ash concrete. Constr Building Mater 20(9):744–753

Kim Y, Hanif A, Usman M, Munir MJ, Kazmi SMS, Kim S (2018) Slag waste incorporation in high early strength concrete as cement replacement: environmental impact and influence on hydration & durability attributes. J Clean Prod 172:3056–3065. https://doi.org/10.1016/j.jclepro.2017.11.105

Long WJ, Zheng D, Duan HB, Han N, Xing F (2018) Performance enhancement and environmental impact of cement composites containing graphene oxide with recycled fine aggregates. J Clean Prod 194:193–202. https://doi.org/10.1016/j.jclepro.2018.05.108

Lúxan MP, Madruga F, Saavedra J (1989) Rapid evaluation of pozzolanic activity of natural products by conductivity measurement. Cem Concr Res 19:63–68. https://doi.org/10.1016/0008-8846(89)90066-5

Medeiros MHF (2002) Concrete structures with corrosion of reinforcements by carbonation: comparison of repair mortars for steel protection. Dissertação (mestrado)—Universidade de São Paulo

Medeiros MHF (2008) contribution to the study of concrete durability with surface retention against the action of chloride ions. Tese (Doutorado)—Universidade de São Paulo

Medeiros MHF, Filho JH, Helene P (2009) Influence of the slice position on chloride migration tests for concrete in marine conditions. Mar Struct 22:128–141. https://doi.org/10.1016/j.marstruc.2008.09.003

Mehta PK, Monteiro PJM (2014) Concrete: microstructure. Properties and Materials Properties and Materials, IBRACON

Miller SA, Myers RJ (2020) Environmental impacts of alternative cement binders. Environ Sci Technol 54:677–686. https://doi.org/10.1021/acs.est.9b05550

Molin Filho RGD, Longhi DA, Souza RCT, Vanderlei RD, Paraíso PR, Jorge LMM (2019) Study of the compressive and tensile strenghts of self-compacting concrete with sugarcane bagasse ash. Rev. IBRACON Estruturas e Mater, 12, 874–883. https://doi.org/10.1590/s1983-41952019000400009

Moretti JP, Nunes S, Sales A (2018) Self-compacting concrete incorporating sugarcane bagasse ash. Constr Build Mater 172:635–649. https://doi.org/10.1016/j.conbuildmat.2018.03.277

Moumin G, Ryssel M, Zhao L, Markewitz P, Sattler C, Robinius M, Stolten D (2020) CO2 emission reduction in the cement industry by using a solar calciner. Renew Energy 145:1578–1596. https://doi.org/10.1016/j.renene.2019.07.045

NBR NM 18 (2012) Portland cement—Chemical anlysis—Determination of loss on ignition Available online: https://www.abntcatalogo.com.br/norma.aspx?ID=91002. Accessed Jun 17, 2020

NBR NM 45—Aggregates—Determination of unit mass and volume of voids. Rio de Janeiro (2006)

Neville AM (2015) Properties of concrete, 5th edn. São Paulo, Bookman

Nguyen T, Mai H, Phan D, Nguyen D (2020) responses of concrete using steel slag as coarse aggregate replacement under splitting and flexure, 1–22

Parra C, Sánchez EM, Miñano I, Benito F, Hidalgo P (2019) Recycled plastic and cork waste for structural lightweight concrete production. Sustain 11:1–15. https://doi.org/10.3390/su11071876

Payá J (2001) Enhanced conductivity measurement techniques for evaluation of fly ash pozzolanic activity. Cemente Concrete Res 31(1):41–49

Payá J (2002) Studies in its properties for reusing in concrete production. J Chem Tecnol Biotechnol 77:321–325

Payá J, Borrachero M, Monzó J, Peris-Mora E, Amahjour F (2001a) Enhanced conductivity measurement techniques for evaluation of fly ash pozzolanic activity. Cem Concr Res 31:41–49. https://doi.org/10.1016/S0008-8846(00)00434-8

Payá J, Monzó J, Borrachero MV, Mellado A, Ordonez LM (2001b) Determination of amorphous silica in rice husk ash by a rapid analytical method. Cement Concrete Res 31:227–231

Payá J, Monzo J, Borrachero MV et al (2002) Sugar-Cane Bagasse Ash (SCBA): Studies on Its roperties for Reusing in Concrete roduction. J Chem Technol Biotechnol 77(3):321–325

Pushkar S (2019) The effect of different concrete designs on the life-cycle assessment of the environmental impacts of concretes containing furnace bottom-ash instead of sand. Sustain, 11. https://doi.org/10.3390/su11154083

Raverdy M, Brivot F, Paillere AM, Bron R (1980) Appréciation de l'activité pouzzolanique de constituents secondaires. In: Proceedings of the 7e Congrés International de la Chimie des Ciments; Paris, pp 36–41

Ribeiro DV, Morelli MR (2014) Effect of Calcination temperature on the pozzolanic activity of Brazilian sugar cane bagasse Ash (SCBA). Mater Res (São Carlos. Impresso) 17: 974981

Rilem Recommendations (1988). CPC-18. Measurement of hardened concrete carbonation depth

Rodolpho PM, Isa MM (1999) A Influência da Resistência à Penetração de Cloretos e da Carbonatação na Durabilidade das Construções de Microconcreto. In: CONGRESO DE CONTROL DE CALIDAD, 7, Montevideo. Anais. Montevideo, 2v. v.1., pp 313–320

Ruan S, Unluer C (2017) Influence of supplementary cementitious materials on the performance and environmental impacts of reactive magnesia cement concrete. J Clean Prod 159:62–73. https://doi.org/10.1016/j.jclepro.2017.05.044

Rukzon S, Chindaprasirt P (2010) Strength and carbonation model of rice husk ash cement mortar with different fineness. J Mater Civil Eng 22(3):253–259

Saffari A, Ataei M, Sereshki F, Naderi M (2019) Environmental impact assessment (EIA) by using the Fuzzy Delphi Folchi (FDF) method (case study: Shahrood cement plant, Iran). Environ Dev Sustain 21:817–860. https://doi.org/10.1007/s10668-017-0063-1

Sagastume Gutiérrez A, Cabello Eras JJ, Gaviria CA, Van Caneghem J, Vandecasteele C (2017) Improved selection of the functional unit in environmental impact assessment of cement. J Clean Prod 168:463–473. https://doi.org/10.1016/j.jclepro.2017.09.007

Sakir S, Raman SN, Safiuddin M, Amrul Kaish ABM, Mutalib AA (2020) Utilization of by-products and wastes as supplementary cementitious materials in structural mortar for sustainable construction. Sustain, 12. https://doi.org/10.3390/su12093888

Sánchez Berriel S, Favier A, Rosa Domínguez E, Sánchez MacHado IR, Heierli U, Scrivener K, Martirena Hernández F, Habert G (2016) Assessing the environmental and economic potential of Limestone Calcined Clay Cement in Cuba. J Clean Prod 124:361–369. https://doi.org/10.1016/j.jclepro.2016.02.125

Santos TA (2017) Evaluation of the calorific potential of sugarcane bagasse and the pozzolanic activity of the ashes generated. MSc Thesis—Universidade Federal da Bahia. Salvador, Brazil. (in Portuguese)

Shen W, Liu Y, Yan B, Wang J, He P, Zhou C, Huo X, Zhang W, Xu G, Ding Q (2017) Cement industry of China: driving force, environment impact and sustainable development. Renew Sustain Energy Rev 75:618–628. https://doi.org/10.1016/j.rser.2016.11.033

Shirkhani A, Kouchaki-Penchah H, Azmoodeh-Mishamandani A (2018) Environmental and exergetic impacts of cement production: a case study. Environ Prog Sustain Energy 37:2042–2049. https://doi.org/10.1002/ep.12894

Silva EJ (2016) Utilização do método de condutividade elétrica para análise da pozolanicidade da cinza do bagaço de cana de açúcar. Dissertação (mestrado)—Universidade Federal de Pernambuco. Caruaru

Singh M, Choudhary K, Srivastava A, Singh Sangwan K, Bhunia D (2017) A study on environmental and economic impacts of using waste marble powder in concrete. J Build Eng 13:87–95. https://doi.org/10.1016/j.jobe.2017.07.009

Singh S, Ransinchung GD, Debbarma S, Kumar P (2018) Utilization of reclaimed asphalt pavement aggregates containing waste from Sugarcane Mill for production of concrete mixes. J Clean Prod 174:42–52. https://doi.org/10.1016/j.jclepro.2017.10.179

Sitter WR (1984) Costs for Service Life Optimization. The "law of fives". In: CEB-RILEM Durability of concrete structures. Proceedings of the International Workshop held in Copenhagem on 18–20 May, 1984

Sofi M, Sabri Y, Zhou Z, Mendis P (2019) Transforming municipal solid waste into construction materials. Sustain 11:1–22. https://doi.org/10.3390/su11092661

Song D, Yang J, Chen B, Hayat T, Alsaedi A (2016) Life-cycle environmental impact analysis of a typical cement production chain. Appl Energy 164:916–923. https://doi.org/10.1016/j.apenergy.2015.09.003

Sperbe JEK (2003) Carbonation and capillary absorption of concretes with mineral additions and chemical activators. MSc Thesis, Universidade Federal de Santa Maria, Santa Maria, Brazil

Swamy RN, Suryavanshi AK, Tanikawa S (1998) Protective ability of an acrylic-based surface coating system against chloride and carbonation penetration into concrete. ACI Mater J 95(2):101–112

Szulej J, Ogrodnik P, Klimek B (2019) Zeolite tuff and recycled ceramic sanitary ware aggregate in production of concrete. Sustain 11:1–16. https://doi.org/10.3390/su11061782

Taylor WHF (1998) Cement chemistry, 2nd edn. Thomas Telford, London

Turk K, Karatas M, Gonen T (2013) Effect of fly ash and silica fume on compressive strength, sorptivity and carbonation of SCC. KSCE J Civil Eng 17(1):202–209

Vázquez-Rowe I, Ziegler-Rodriguez K, Laso J, Quispe I, Aldaco R, Kahhat R (2019) Production of cement in Peru: Understanding carbon-related environmental impacts and their policy implications. Resour Conserv Recycl 142:283–292. https://doi.org/10.1016/j.resconrec.2018.12.017

Villarino BL, Sevilla JB, Luque VF, García MRR (1995) Concrete durability. Influence of chloride ions concentrations in mixing water and dissolution. Materiales de Construcción 45(237):37–52

Warati GK, Darwish MM, Feyessa FF, Ghebrab T (2019) Suitability of scoria as fine aggregate and its effect on the properties of concrete. Sustain, 11. https://doi.org/10.3390/su11174647

Wilk E, Krówczyńska M, Zagajewski B (2019) Modelling the spatial distribution of asbestos-cement products in Poland with the use of the random forest algorithm. Sustain, 11. https://doi.org/10.3390/su11164355
Wu M, Zhang Y, Ji Y, Liu G, Liu C, She W, Sun W (2018) Reducing environmental impacts and carbon emissions: Study of effects of superfine cement particles on blended cement containing high volume mineral admixtures. J Clean Prod 196:358–369. https://doi.org/10.1016/j.jclepro.2018.06.079
Xu P, Zhao Q, Qiu W, Xue Y, Li N (2019) Microstructure and strength of Alkali-activated bricks containing municipal solid waste incineration (MSWI) fly ash developed as construction materials. Sustain, 11, https://doi.org/10.3390/su11051283
Younsi A, Davson C et al (2013) Accelerated carbonation of concrete with high content of mineral additions: effect of interactions between hydration and drying. Cement Concrete Res 43:25–33

The manufacturer's authorised representative in the EU is Springer
Nature Customer Service Centre GmbH, Europaplatz 3, 69115 Heidelberg,
Germany. If you have any concerns regarding our products, please
contact ProductSafety@springernature.com

Printed and bound by CPI Group (UK) Ltd, Croydon, CR0 4YY
28/09/2026
02224086-0020